Make:
Fire

Make:
Fire

The Art and Science of Working with Propane

TIM DEAGAN

Make: Fire
The Art and Science of Working with Propane
By Tim Deagan

Published by Maker Media, Inc.,
150 Todd Road, Suite 100
Santa Rosa, California 95407.

Maker Media books may be purchased for educational, business, or sales promotional use. Online editions are also available for most titles (*oreilly.com*). For more information, contact our corporate/institutional sales department: 800-998-9938 or *corporate@oreilly.com*.

Publisher: Roger Stewart
Editor: Roger Stewart
Copy Editor: Elizabeth Campbell, Happenstance Type-O-Rama
Technical Reviewer: Dave X
Proofreader: Elizabeth Welch, Happenstance Type-O-Rama
Interior Designer and Compositor: Maureen Forys, Happenstance Type-O-Rama
Cover Designer: Maureen Forys, Happenstance Type-O-Rama
Indexer: Valerie Perry, Happenstance Type-O-Rama

Photo credits
Chapter 1 opening photo: Hep Svada, 2015 Maker Faire Bay Area, Angel of the Apocalypse sculpture by Flaming Lotus Girls

Chapter 2 opening photo: Hep Svada, 2015 Maker Faire Bay Area, Chester "The Fire Breathing Horse Artcar" designed by Jason Anderholm, Rebecca Anders, and Don Cain

Chapter 13 opening photo: Caroline Mills (mills), 2008 Burning Man, "Mutopia," The Flaming Lotus Girls

April 2016: First Edition

Revision History for the First Edition
2016-04-15: First Release (TI)
2023-07-28: Second Release (LSI)
See *oreilly.com/catalog/errata.csp?isbn=9781680450873* for release details.

978-1-680-45087-3

Safari® Books Online

Safari Books Online is an on-demand digital library that delivers expert content in both book and video form from the world's leading authors in technology and business.

Technology professionals, software developers, web designers, and business and creative professionals use Safari Books Online as their primary resource for research, problem solving, learning, and certification training.

Safari Books Online offers a range of plans and pricing for enterprise, government, education, and individuals. Members have access to thousands of books, training videos, and prepublication manuscripts in one fully searchable database from publishers like O'Reilly Media, Prentice Hall Professional, Addison-Wesley Professional, Microsoft Press, Sams, Que, Peachpit Press, Focal Press, Cisco Press, John Wiley & Sons, Syngress, Morgan Kaufmann, IBM Redbooks, Packt, Adobe Press, FT Press, Apress, Manning, New Riders, McGraw-Hill, Jones & Bartlett, Course Technology, and hundreds more. For more information about Safari Books Online, please visit us online.

How to Contact Us

Please address comments and questions concerning this book to the publisher:

Make:
150 Todd Road, Suite 100
Santa Rosa, CA 95407
877-306-6253 (in the United States or Canada)
707-639-1355 (international or local)

Make: unites, inspires, informs, and entertains a growing community of resourceful people who undertake amazing projects in their backyards, basements, and garages. Make: celebrates your right to tweak, hack, and bend any technology to your will. The Make: audience continues to be a growing culture and community that believes in bettering ourselves, our environment, our educational system—our entire world. This is much more than an audience, it's a worldwide movement that Make is leading we call it the Maker Movement.

For more information about Make:, visit us online:

- Make: magazine makezine.com/magazine
- Maker Faire makerfaire.com
- Makezine.com makezine.com
- Maker Shed makershed.com

To comment or ask technical questions about this book, send email to *books@make.co*.

To my wife Tracy and daughter Val, who are the brightest fires in my life.

Contents

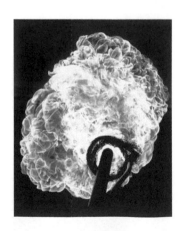

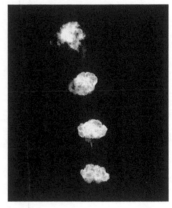

Preface

"Sweet lady propane."
—HANK HILL, *King of the Hill*

AS A KID, I loved building models. Seeing a completed fighter plane, PT boat, or hot rod emerge from injection molded sheets was a major thrill. But by middle school, I found that I was gluing airplane parts to hot wheels cars, mixing boxes of parts to make new models, and generally painting outside the lines. As much fun as it was to build the designers' models, there was an extra thrill in building my own creations.

By high school, I began expanding my toolset and building things that weren't from kits. Line-following and light-sensing robots wandered around my room. I began buying project books and devouring the skills and knowledge they contained. I did build things from the books, but usually just enough to gain some level of mastery. Then I was off on my own again, trying to build things I'd never seen before.

This book is written so that the projects are heavily documented and is designed to leave very little ambiguity about their construction. I've always hated project books that left out important details or were vague about how to do something. I wanted this book to tell you everything you need to know to be successful.

However, this doesn't mean that you should feel you have to slavishly follow these instructions to have fun and build a successful project. There are many ways to accomplish the same thing when it comes to things like plumbing fittings. Sometimes you have to make do with what

GASOL

TRADE MARK

NATIONAL PROPANE GAS ASSOCIATION

you can find. Adapting (safely) is fine. I'll also offer enhancement suggestions with each project. Please be sure to read Chapter 1, "Understanding Propane," so that you have a good foundation in how to work safely with propane. Then modifying the projects to pursue your own goals is perfectly appropriate.

Propane is a wonderfully common fuel source, yet there is a lot of misinformation about it among makers and artists who would like to incorporate it into their work. Blacksmiths and metal casters use propane to power torches; artists and makers use it to create flame effects like flambeaux, booshes, Rubens' tubes, and other visual displays. It's available almost everywhere, is nontoxic and, given simple precautions, very safe to handle. Nevertheless, newcomers often have trouble finding out the requirements for working safely and confidently with propane. This book is an attempt to help people get up to speed on using propane in their projects and to offer some complete DIY projects to start out with.

Despite my trying to write a complete introduction to this topic, let me encourage you to take a class if offered, apprentice with a group if possible, and spend as much time as you can to learn from knowledgeable folks in this field. No book is a full substitute for hands-on experience, but I've tried to provide a good start.

The equipment and safety procedures described in this book are valid for working with propane vapor, not propane liquid; handling propane liquid is beyond the scope of this text.

My apologies to international readers, but it's also important to note that, while the principles are universal, many of the specifics are appropriate only for the United States. Where I feel confident that I am making a valid comparison, I have included either a direct metric conversion of a measurement or a reasonable equivalent that is more likely to be available. In some cases, the variety of international standards and parts is beyond my knowledge. Where this is the case, I have not listed a conversion or metric alternative part. I would rather rely on readers' local knowledge than reference some part that is not safe.

I hope you enjoy this book and go on to create your own amazing flame effects and fire art!

—TIM DEAGAN

What to Expect from This Book

After reading this book you will not be:

- A qualified plumber
- A qualified liquid petroleum gas (LPG) professional

Those professions require training and licensure. Don't underestimate what it takes to learn what these individuals know! If you're lucky enough to know someone who's a licensed plumber or LPG professional, make them cookies or do something to make it worth their while to look over your shoulder as you work with propane!

After reading this book you should be able to:

- Understand how to safely handle propane
- Examine basic propane systems, recognize their parts, and understand how they work
- Identify major safety issues with basic propane systems
- Understand the issues involved with safely constructing and operating basic propane systems

Propane's History

Ever been angry at the price of gasoline? If so, you had the same complaint that led to the discovery of propane! The history of propane can be traced back to an angry consumer in early 1910. Every time he filled the gas tank of his new Model T Ford, half the gasoline was evaporating by the time he got home. He was spitting mad and wanted the government to do something!

The angry Ford owner approached Dr. Walter O. Snelling, a chemist working for the Bureau of Mines. (See Figure 1.) Dr. Snelling was already famous for his underwater explosives igniter, which had saved the United States millions of dollars in constructing the Panama Canal. He agreed to look into the motorist's problem. He filled up a glass jug from the gasoline in the car's tank and couldn't help but notice that, as he took it back to his lab, the stopper kept blowing out of the jug.

FIGURE 1: Dr. Walter O. Snelling, father of modern propane. PUBLIC DOMAIN

Pulling parts from a hot water heater and various lab gear, Dr. Snelling built a still and began distilling, or fractionating, the gasoline. He quickly discovered that the evaporating gases consisted of propane, butane, and other hydrocarbons. The gasoline on the market in those days was what would be referred to as wild gas today. That's gasoline that's pretty much pulled directly out of the well without any processing. Much of the reason our gas is no longer wild is because Dr. Snelling tamed it!

Dr. Snelling soon formed the American Gasol company, and by 1912, propane was heating homes; by 1913, cutting and welding metal; and by 1915, powering cars. The industry never looked back. Dr. Snelling sold his patents, and eventually the company, to Phillips Petroleum, now part of today's ConocoPhillips.

As TV's Hank Hill proclaims, Dr. Snelling was definitely the "father of modern propane."

Propane in the Twenty-First Century

Today, propane is produced as a byproduct of natural gas processing and petroleum refining. About 97% of the propane in the United States comes from North America. Since propane is a byproduct, its production ebbs and flows with the sources it comes from rather than having its own market. Ironically, in 2014, new North American shale gas production generated such large amounts of propane that the export market expanded enough to cause shortages in the United States.

After the North American propane is collected, it's shipped overseas or pressurized to a liquid and stored in giant salt caverns capable of holding 80 million barrels. These are located at Fort Saskatchewan, Alberta; Mont Belvieu, Texas; and Conway, Kansas.

Propane is sold in different grades. These grades represent the percentage of propane and the other allowed components. Consumer grade, known as heavy-duty 5% (HD-5), requires a minimum of 90% propane and allows up to 5% propylene and 5% butane, ethane, and methane, as well as odorants such as ethyl mercaptan. Two other lesser grades are known as HD-10 and commercial grade. In some parts of the world, products labeled as propane may be up to 50% butane.

Tens of millions of American customers use propane yearly. In 2012, the United States produced about 15 billion gallons of propane for use in grills, home heating, transportation fuel, and industrial activities. This is expected to rise as the world hunger for fuel increases.

Propane is considered an alternative fuel in the United States due to its domestic production, clean burning, and relatively low cost. It's the third most popular transportation fuel after gas and diesel. In 2013, it was only 2% of the energy usage in the United States and only 2% of that was used in transportation, so there's a long way to go. Nevertheless, as a fossil fuel, it's not a renewable resource.

Propane also shows up in some unexpected places. Due to its lack of toxicity and relative inertness, propane has been re-branded green gas and is used, when combined with silicon oil, as a propellant for paintball guns. Propane, and its brother butane, have become the most common replacement for chlorofluorocarbons as propellants in aerosol cans. Any place that a nontoxic pressurized gas source is needed, propane is a likely candidate.

More Online

Visit *http://makefirebook.com* for PDFs of the block diagrams and more amazing images.

Acknowledgments

I'M VERY LUCKY TO write this book and couldn't have done so without tremendous help. However, any mistakes in this book are my own and despite the best efforts of those who helped me. But I must say thank you to a few people. My wonderful wife Tracy gifted me the space for me to fill our house and yard and lives with tools and projects. My daughter Val makes me want to leave something for the future. My editor, Roger Stewart, made it possible for my ideas to be published. My copyeditor, Elizabeth Campbell of Happenstance Type-O-Rama, made my writing readable.

The wonderful freaks of Burning Flipside created a community where I could learn and play. The guys at Harrell's Hardware in Austin, one of the last and best of the neighborhood hardware stores, helped me solve innumerable problems (their motto; "Together we can do it yourself!"). Ron Reil, a maker hero of mine for years, became a guide to me and made the burner chapter possible. DaveX, the manager of the Fire Safety Team at Burning Man for over 15 years, helped me understand a number of the pitfalls and requirements of building flame effects. Bob, Bean, and Hunter supported me with love and space to test my fire. Jeremy "Chainsaw" Williams and David McGriffy provided critical sounding boards. The Flaming Lotus Girls offered me insight, enthusiasm, and a level of build that I'll always aspire to. Michael Boyd provided me with new ideas. Capt. Stacey Cox and his team at the Austin Fire Department's Special Events Division kept me on the straight and narrow, helping me understand the practical aspects of permitting and public displays of flame effects. Kami Wilt and the crew of the Austin Mini Maker Faire gave me their trust and belief that we need to give more back to the maker community. Steve Wolf of Wolf Stuntworks and Stunt Ranch imparted momentum to me that got me writing. Lee Thomas and the rest of the wonderful people at CORT Business Services kept me gainfully employed and able to pay my bills. Thank you all and know that this only scratches the surface of the debts I owe in writing this book.

—Tim Deagan,
Austin, TX

About the Author

TIM DEAGAN likes to make things. He casts, prints, screens, welds, brazes, bends, screws, glues, nails, and dreams in his Austin, Texas, shop. He's spent decades gathering tools based on the idea that one day he will come up with a project that has a special use for each and every one of them.

Tim likes to learn and try new things. A career troubleshooter, he designs, writes, and debugs code to pay the bills. He has worked as a stagehand, meat cutter, speechwriter, programmer, sales associate at Radio Shack, VJ, sandwich maker, computer

tech support specialist, car washer, desk clerk, DBA, virtual CIO, and technical writer. He's run archeology field labs, darkrooms, produce teams, video stores, ice cream shops, consulting teams, developers, and QA teams. He's written for *Make:* magazine, *Nuts & Volts*, *Lotus Notes Advisor*, and *Databased Advisor* magazines.

Tim collects board games, Little Mermaid stuff, ukuleles, accordions, tools, watches, slide rules, graphic novels, art supplies, hobbies, books, gadgets, and sharp and pointy things. He owned, and escaped from owning, a 1960 Ford C-850 Young Fire Equipment fire engine (though he kept the siren). Tim paints, sketches, sculpts, quilts, sews, and works leather. Tim has climbed antenna towers, wrecked motorcycles, learned to parasail, and jumped out of perfectly good airplanes,

Tim has been or is a boy scout, altar boy, Red Cross disaster action team captain, volunteer firefighter, flyman, Wocista, Flipside burner, actor, Austin Mini Maker Faire flame and safety coordinator, lighting tech, ham radio operator (KC5QFG), musician, and licensed Texas flame effect operator. Tim has studied Daito Ryu AikiJujitsu with Sensei Rick Fine, and Tomiki Aikido with Sensei Strange.

Tim loves his wife, his daughter, his dogs, and his friends, and feels very lucky indeed to be able to write all the lists above.

Understanding Propane

WHEN I'M INSPECTING PEOPLE'S propane projects as a *flame effect operator* (FEO; the licensed person at an event where fire is involved), or as the flame and safety coordinator at the Austin Mini Maker Faire, I frequently end up trying to explain to people why their project, device, or effect is unsafe. Most of the time, it's because they have no basic understanding of what propane is or how it interacts with the world.

Understanding the fundamentals of propane will make everything you do with it safer and easier. You don't need extensive chemistry or engineering knowledge. This book will provide you with everything you need to know to make use of this amazingly useful fuel. Once you've read this and the following chapter, you'll be able to look at propane projects and understand with confidence what's safe and what's not.

I cannot promise you that people won't argue with you or reject my standards of safety if they have built to different standards. I hope you will use good judgment and assess the situation based on basic principles. It's a valid critique that what I share in this book is a conservative approach, but that's because I want you to stay safe and have fun.

Basic Propane Chemistry

Let's start with the fundamentals. What, exactly, is propane? Like a lot of other interesting things, it's a hydrocarbon. That means it's made of nothing but hydrogen and . . . you guessed it, carbon. Pretty basic stuff.

There are 134 (as of 2015) synonyms listed for propane in the PubChem database but the most common are:

- C3H8
- CH3CH2CH3
- n-Propane
- propan
- Dimethylmethane
- Propyldihydride
- Propyl hydride
- Petroleum gas, liquefied
- Hydrocarbon Propellant A-108

The term *propane* is often used for propane vapor. Liquid propane is usually referred to as liquefied petroleum gas (LPG). The terms *LPG* and *propane* are occasionally used interchangeably, but this text is focused on propane vapor so, for the most part, we'll avoid using the term LPG.

The backbone of a hydrocarbon is a carbon skeleton. This frame is adorned with hydrogen atoms sticking out as far away from each other as they can (their electrons find one another repellant). Most fuels we use are hydrocarbons. Propane has three carbon atoms and eight hydrogen atoms. The carbons are arranged in a zigzag chain, with hydrogen atoms sticking out like antennae. (See Figure 1-1.)

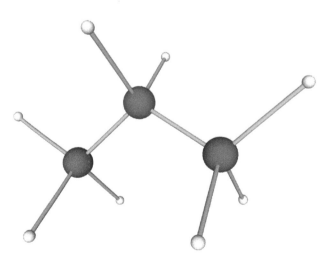

FIGURE 1-1: **Our friend, the propane molecule**

Other collections of carbon and hydrogen atoms that share this structural pattern are called linear alkanes. They include methane, ethane, butane, pentane, hexane, and more. (See Figure 1-2.) Methane is the simplest with one carbon atom; the rest are the result of adding more. Methane is so common that it's believed to be present on every planet in the solar system as well as the major moons!

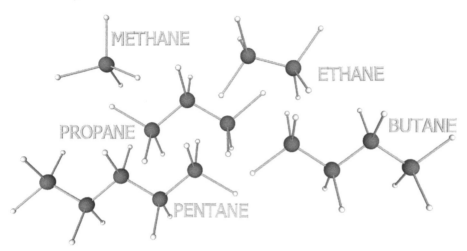

FIGURE 1-2: **Propane and related alkanes**

Properties of Propane

Propane, by itself, is not a greenhouse gas. For one thing, it's heavier than air so it doesn't rise into the upper atmosphere. However, many of the resultant products from burning it are greenhouse gases. Carbon dioxide and various combinations of nitrogen and oxygen that result from incomplete combustion are greenhouse gases. In this regard, propane is considered a better fuel than its alternatives, so it gets reasonably good marks from the conservation community.

Chemically, propane is extremely safe. It's completely non-toxic and noncarcinogenic, even with exposure over time. It's colorless and odorless. If you think you smell propane, you're really smelling an additive like ethyl mercaptan, put there so you can detect the presence of leaked propane vapor.

This should not be interpreted to say that propane isn't dangerous. It's just that the danger is fundamentally mechanical (and it's extremely dangerous in that respect). It's the pressures and temperatures that are involved that constitute propane's risks. Propane accidents can potentially cause explosions, burns, frostbite, and asphyxiation. But compared to other fuels, these accidents are relatively easy to avoid with safe practices. We'll look at the specifics of handling propane safely in later chapters. For now, understanding the properties of propane will help those practices make sense.

At normal pressures, propane boils at −44°F (−42.2°C). So unless you're in deeply arctic conditions, you will only encounter unpressurized propane as vapor. Inside the cylinder at room temperature, the propane boils until it fills the empty space (aka, *headspace*) with enough gas to pressurize the system and keep the rest of the propane liquid. (See Figure 1-3.)

At normal temperatures and pressure, propane liquid will expand 270 times to become vapor. This means that one gallon of propane will become 36 cubic feet (1.02 cubic meters) of vapor if released from its container. (See Figure 1-4.) However, that vapor will mix with air (we'll learn later about the propane-air mixture level that allows propane to combust, but it's around 5% propane and 95% air) so that the amount of combustible vapor is much larger than the propane vapor alone. (See Figure 1-5.)

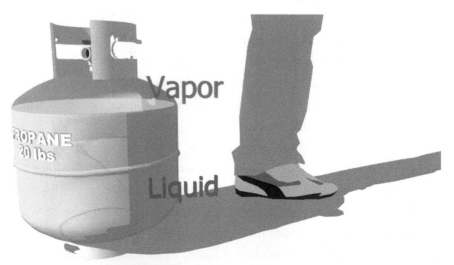

FIGURE 1-3: A propane cylinder contains liquid and vapor.

FIGURE 1-4: Comparing unpressurized vapor to cylinder sizes

FIGURE 1-5: Mixed with air, the combustible mixture is 20 times larger than propane vapor alone.

Propane vapor is one and a half times heavier than air. This causes it to sink rather than rise. **This is important.** If an indoor propane system leaks, the propane will pool in the bottom of the room, potentially asphyxiating people or causing an explosive hazard as it mixes with the air.

Propane is lighter than water. A gallon of water weighs 8.33 pounds (1 L = 1 kg). A gallon of propane weighs 4.24 pounds (1 L = 1.93 kg).

Propane has a higher octane rating than gasoline. Most gasoline has an octane rating of 87–91; propane's rating is 110. However, octane only really measures how much compression the fuel can withstand before igniting. Propane actually has a lower energy density than gasoline. This is typically measured in British Thermal Units (BTU). A gallon of gas can produce 125,000 BTUs; a gallon of propane can produce 91,700 BTUs. Sometimes this is expressed as kilowatt hours per gallon (kWh/g). Propane has 26.8 kWh/gallon; gasoline has 36.6 kWh/gallon.

Liquid propane is an excellent solvent of petroleum fractions, vegetable oils and fats, natural rubber and organic compounds of sulfur, oxygen, and nitrogen. Acetylene red welding hose and other natural rubber hose is not appropriate for propane use due to its composition, nor is any equipment with rubber O-rings or seals.

Propane does not corrode or dissolve metals, polyvinyl chloride (PVC), or polyethylene (PE). This does not mean that these are necessarily acceptable materials for piping or containing propane. Much as with humans, the threat is typically not chemical. It's the mechanical properties of propane relating to pressure and temperature that may cause these materials to fail (perhaps catastrophically). In later chapters, we will discuss the appropriate materials for working with propane based on various legal codes.

Propane and Natural Gas

It's worth taking a moment to talk about the differences between propane and natural gas. Many American homes have natural gas piped in to heat furnaces, hot water heaters, and clothes dryers.

Natural gas, as recovered out of the ground, is a mix of different gases.

The refinement of natural gas is one of the primary means of acquiring propane. After refinement, the natural gas delivered to your home is almost entirely methane. Unlike propane, natural gas is lighter than air; it will rise rather than sink.

TABLE 1-1: Contents of natural gas

COMPONENT	SYMBOL	%
Methane	CH4	70–90%
Ethane	C2H6	0–20%
Propane	C3H8	0–20%
Butane	C4H10	0–20%
Carbon Dioxide	CO2	0–8%
Oxygen	O2	0–0.2%
Nitrogen	N2	0–5%
Hydrogen Sulphide	H2S	0–5%
Rare gases	A, He, Ne, Xe	trace

Natural gas has an octane rating of 130, and has a higher energy density than propane. A cubic foot of propane will produce 1,030 BTUs; a cubic foot of natural gas, 2,516 BTUs. Natural gas also burns in a slightly different concentration of air, 5–15%, than propane.

Pressure

Pounds per square inch (psi) is a unit of measure. Standard atmospheric pressure is 14.7 psi (101325 Pa). *Pounds per square inch absolute (psia)* is the measure of pressure relative to full vacuum. *Pounds per square inch gage (psig)* is the pressure relative to the ambient atmosphere. When you fill a tire with 35psi that is really 35 psig because you're adding 35 psi above the atmospheric pressure. Many gauges that are labelled psi are actually psig and common use of psi is really referring to psig. Therefore, unless I specify otherwise if I use the term psi in this book, I'll be referring to psig.

Understanding how propane and pressure interrelate is critical to working safely. The most important fundamental principle is that **pressure and temperature are joined at the hip**. Any change to one will guarantee a change in the other in the same direction. Raising the temperature will raise the pressure. Lowering the pressure will lower the temperature.

Liquid volume is, for all practical purposes, unrelated to pressure. A 1lb bottle of camping propane and a 500 gallon propane tank are under exactly the same pressure if they're the same temperature. (See Figure 1-6.) From the perspective of pressure, all propane containers require equal caution.

If the Temperature
is the same, the Pressure
is the same in all these
containers

Propane

FIGURE 1-6: Pressure is the same in all cylinders at the same temp

So, what's the pressure in a propane cylinder? We just came to the understanding that the size of the cylinder doesn't matter. The thing you need to know to answer the question is . . . temperature! At 32°F (0°C) the pressure is approximately 45 psig (60 psi). At 72°F (22.2°C) the pressure climbs to 115 psig (130 psi). A nearly empty cylinder and a completely full one maintain the same pressure if the temperature is the same. (See Figure 1-7.)

To understand exactly why, we'd have to examine the three most basic laws governing the relationship between temperature, pressure, and volume of gases. These are Boyle's Law, Charles's Law, and Gay-Lussac's Law. I'll cover these briefly in Appendix A and provide links to learn more online if you're interested.

For our purposes, since the volume of the gas is constrained by the cylinder, the relationship between temperature and pressure is the one that matters the most to us.

The ambient air temperature isn't really the temperature we're interested in. The cylinder temperature may start at the ambient air temperature, but if you draw propane from the cylinder, you reduce the pressure (think of letting air out of a tire).

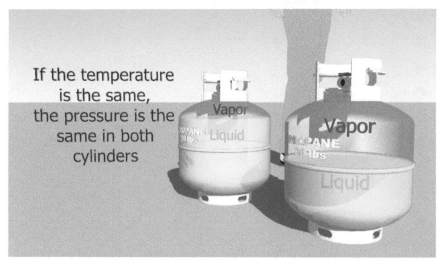

FIGURE 1-7: Pressure is the same in cylinders at the same temperature, regardless of how full they are.

The boiling point of liquids lowers as the pressure decreases (and rises as it increases). This is why water boils at a lower temperature as the altitude increases (and air pressure drops). A really weird experiment is to put water at a temperature below 32°F (0°C) in a bell jar and reduce the pressure until it's freezing and boiling at the same time!

The following chart shows the boiling point of propane at different pressures. This is also known as a vapor pressure chart or vapor pressure curve (see Figure 1-8).

TABLE 1-2: Propane vapor pressure

°F	°C	PSIG
<= −45	<= −43	0
−44	−42	Boiling begins at sea level
−20	−29	11
−10	−23	16
0	−18	24

(continues)

(continued)

°F	°C	PSIG
10	−12	32
20	−7	41
30	−1	51
40	4	64
50	10	78
60	16	92
70	21	111
80	27	128
90	32	149
100	38	172
110	43	197
120	49	230
130	54	257

In graphic form this is represented as a curve (see Figure 1-8):

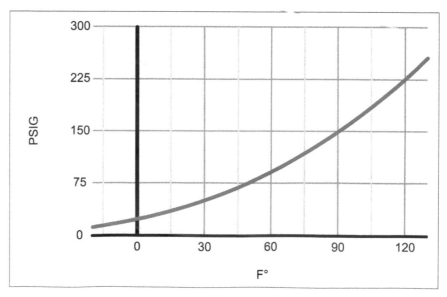

FIGURE 1-8: The propane vapor pressure curve

Relationship between Temperature and Pressure

As long as the liquid propane is above its boiling point, it will boil and fill the cylinder back up with vapor. Energy for boiling has to come from somewhere and in our case it comes from the stored heat in the liquid propane. Using up that stored heat reduces the temperature of the liquid propane (and by conduction, the cylinder itself).

If you keep drawing propane from the cylinder, the pressure (and hence temperature) will keep dropping until the cylinder is coated with ice (from water in the air) and the pressure falls to the point that hardly any propane will come out.

Let's look at a theoretical sequence of steps to illustrate this.

For this exercise, let's say it's a nice cool day in Texas and the ambient air temp is 80°F (27°C). (See Figure 1-9.)

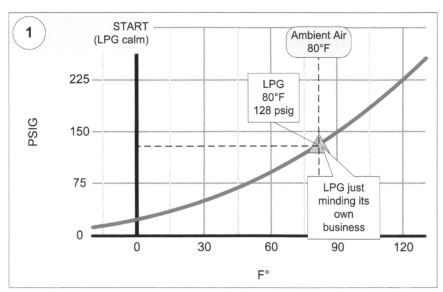

FIGURE 1-9: Temperature pressure sequence step 1

Assuming we're releasing a lot of vapor quickly through a boosh or some other flame effect, we drop the pressure quickly. The liquid propane boils to try and restore the pressure, which takes energy from the latent heat in the propane and it cools down. (See Figure 1-10.)

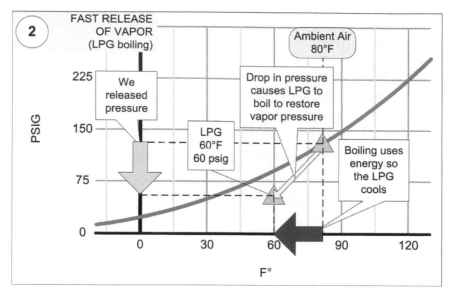

FIGURE 1-10: Temperature pressure sequence step 2

You can release pressure faster than you can draw heat, so you end up with the temperature and pressure in an unstable state (below vapor pressure).

The boiling propane loses heat and raises the pressure in the cylinder until a stable vapor pressure is reached for the current temperature. (See Figure 1-11.)

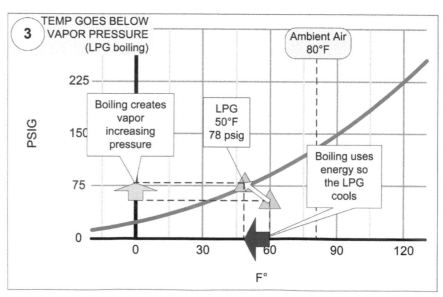

FIGURE 1-11: Temperature pressure sequence step 3

At the point where the temperature and pressure are stable, the liquid propane stops boiling. The maximum pressure available from the cylinder is now considerably lower than it was when we started. (See Figure 1-12.)

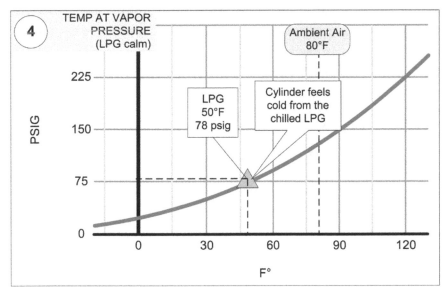

FIGURE 1-12: Temperature pressure sequence step 4

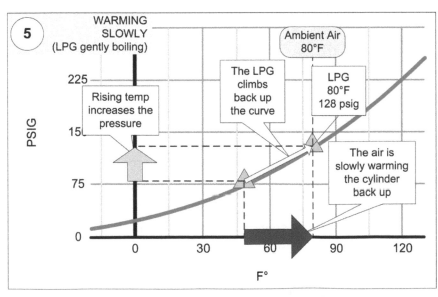

FIGURE 1-13: Temperature pressure sequence step 5

Since the air temperature is now higher than the temperature of the liquid propane, the system begins to warm. This causes repetitions of tiny versions of step 3 where the propane boils a little, the temperature drops a little, the pressure rises a little, the system warms a little, and the temperature and pressure slowly climb the vapor pressure curve until the liquid reaches the temperature of the ambient air. (See Figure 1-13.)

All of which brings us back to where we started. (See Figure 1-14.)

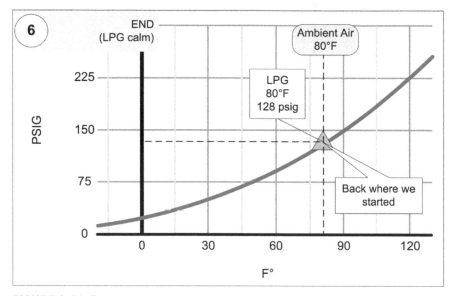

FIGURE 1-14: Temperature pressure sequence step 6

Combustion

For most readers, the point of working with propane is to burn it. Combustion is the general term for burning things. In this section, we'll examine the details of propane combustion.

Propane requires a spark to ignite in air unless the temperature is above 920°F (493°C)–1020°F (549°C). Above those temperatures, it will spontaneously ignite. Even then, propane will only burn when it is in a specific range of propane-air percentages. This is true of all fuel gases (see Figure 1-15) with the lower limit referred to as the *lower explosive limit (LEL)* and the upper as the *upper explosive limit (UEL).* Various texts disagree on the specific values, but in general, propane's LEL is 2.1% and the UEL is somewhere between 9.6% and 10.1%. This variance is probably related to the difference in the amount of butane and other gases present in different grades of propane, which impacts flammability. For propane to burn cleanly, (ie, only producing carbon dioxide and water as byproducts) it must be 4.2% in proportion to oxygen. This is known as complete combustion (or *stoichiometric combustion* for you chemists and word lovers).

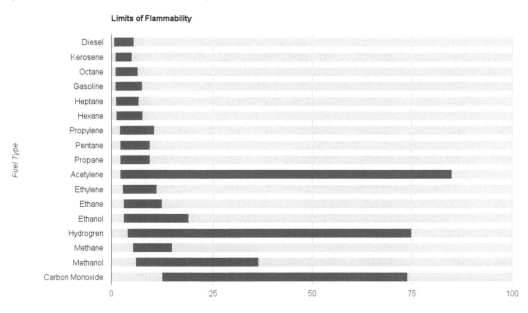

FIGURE 1-15: Limits of flammability for common fuel gases

Burning with less-than-ideal propane results in a lean burn, where the flames lift from the burner and try to go out. This is an oxidizing flame that introduces extra O_2 into the atmosphere.

Burning with more than the ideal percentage of propane is a rich burn. This creates large, yellow flames and possibly soot (which is actually just carbon). This is a reducing flame that will pull oxygen from the environment. This type of incomplete combustion creates carbon monoxide and possibly carbon (the soot).

TABLE 1-3: Propane combustion

% PROPANE IN AIR	COMBUSTION	BURN	REACTION	FLAME	EQUATION
10.50%	Not Flammable				
8.40%	Incomplete	Rich	Reducing	Large Yellow Flame, Soot	$C_3H_8 + 2.5O_2 \rightarrow 4H_2O + CO + 2C$ (+heat, +light)
7.00%	Incomplete	Rich	Reducing	Large Yellow Flame, Soot	$C_3H_8 + 3O_2 \rightarrow 4H_2O + 2CO + 1C$
6.00%	Incomplete	Rich	Reducing	Large Yellow Flame	$C_3H_8 + 3.5O_2 \rightarrow 4H_2O + 3CO$ (+heat, +light)
5.25%	Incomplete	Rich	Reducing	Large Yellow Flame	$C_3H_8 + 4O_2 \rightarrow 4H_2O + CO_2 + 2CO$ (+heat, +light)
4.67%	Incomplete	Rich	Reducing	Large Yellow Flame	$C_3H_8 + 4.5O_2 \rightarrow 4H_2O + 2CO_2 + CO$ (+heat, +light)
4.20%	Complete	Stoichiometric	Neutral	Blue Flame	$C_3H_8 + 5O_2 \rightarrow 4H_2O + 3CO_2$ (+heat, +light)
3.82%	Incomplete	Lean	Oxidizing	Flames lift/go out	$C_3H_8 + 5.5O_2 \rightarrow 4H_2O + 3CO_2 + .5O_2$ (+heat, +light)
3.50%	Incomplete	Lean	Oxidizing	Flames lift/go out	$C_3H_8 + 6O_2 \rightarrow 4H_2O + 3CO_2 + 1O_2$ (+heat +light)
3.23%	Incomplete	Lean	Oxidizing	Flames lift/go out	$C_3H_8 + 6.5O_2 \rightarrow 4H_2O + 3CO_2 + 1.5O_2$ (+heat, +light)
3.00%	Incomplete	Lean	Oxidizing	Flames lift/go out	$C_3H_8 + 7O_2 \rightarrow 4H_2O + 3CO_2 + _2O_2$ (+heat, +light)
2.80%	Incomplete	Lean	Oxidizing	Flames lift/go out	$C_3H_8 + 7.5O_2 \rightarrow 4H_2O + 3CO_2 + 2.5O_2$ (+heat, +light)
2.63%	Incomplete	Lean	Oxidizing	Flames lift/go out	$C_3H_8 + 8O_2 \rightarrow 4H_2O + 3CO_2 + 3O_2$ (+heat, +light)
2.47%	Incomplete	Lean	Oxidizing	Flames lift/go out	$C_3H_8 + 8.5O_2 \rightarrow 4H_2O + 3CO_2 + 3.5O_2$ (+heat, +light)
2.33%	Incomplete	Lean	Oxidizing	Flames lift/go out	$C_3H_8 + 9O_2 \rightarrow 4H_2O + 3CO_2 + 4O_2$ (+heat, +light)
2.21%	Incomplete	Lean	Oxidizing	Flames lift/go out	$C_3H_8 + 9.5O_2 \rightarrow 4H_2O + 3CO_2 + 4.5O_2$ (+heat, +light)
2.10%	Not Flammable				

PROPANE'S LIMITS OF FLAMMABILITY	APPROXIMATE EQUATION FOR AIR
2.1% – ~9.6%	$(O_2 + 3.76 N_2)$

C_3H8	Propane
O_2	Oxygen
H_2O	Water
C	Carbon
CO	Carbon Monoxide
CO_2	Carbon Dioxide

The flame temperature of propane varies with the type of burn, but propane's peak flame temperature is 3614°F (1990°C).

I hope the details about propane encourage you more than intimidate you. The important thing to understand is that, with care, you can safely work with propane. It won't poison you or give you cancer. There's no mystery about when it will catch fire and when it won't. Compared to other fuels, it's remarkably safe. We'll rely on the information in this chapter throughout the book to help us make good decisions and design choices, so feel free to refer back to these pages as necessary.

HEP SVADA

Equipment and Parts

2

PROPANE IS SO COMMON that appropriate parts are easily found almost everywhere in an industrialized country. The projects we'll be building will use standard parts found in big-box hardware stores, local plumbing supply companies, and on the Internet. Understanding which parts are or are not appropriate for propane use, and why, is what we'll be working toward in this chapter.

As noted at the beginning of this book, the focus is on propane vapor systems. As such, the equipment and parts we'll be talking about are for vapor, not liquid. While many forklifts, vehicles, and industrial systems use propane liquid, these systems generally require a different class of equipment and safety procedures.

Cylinders

The terms *cylinder*, *tank*, and *bottle* are all, more or less, interchangeable. Commonly, large permanent vessels are called tanks, portable vessels are called cylinders and small vessels are called bottles. But this is not a hard-and-fast rule.

We're going to focus on "portable" cylinders, (usually considered to be 100 lbs and below,) rather than on permanently installed propane tanks. The principles are pretty much the same, but most artists and makers will be buying or refilling cylinders ranging from 20 to 100 lbs.

Portable propane cylinders are sometimes referred to by volume and sometimes by weight. A 20 lb (~ 9 kg) cylinder is sometimes referred to as a 5-gallon tank. Metric propane cylinders come in sizes like 7 kg, 12 kg, and so on. With apologies to readers using a more rational system of measurement, I'm going to refer to American propane cylinders using their lb denominations.

The mathematically inclined readers may ask, "Why not 21 lbs (5 gallon $\times$ 4.24 lbs/gallon = 21.2 lbs)?" The reason is because propane is sold by weight, not volume. A standard BBQ cylinder is designed to be filled with 20 lbs of propane. Technically the cylinder should have the capacity to hold another 20%, but the law requires that there be room left in the cylinder for expansion. Propane expands 1.5% for every 10°F (5.5°C), so filling a cylinder completely is a recipe for disaster. This is why, since 2002, propane cylinders between 4 lbs and 40 lbs in the United States are now required to have an overfill protection device (OPD) that limits the fill level to 80%.

However, when cylinders are refilled, they will frequently be filled with even less than 80%. You may want to check what you're paying for. The cylinder has a marking around the handle that says something like "TW - 16.6LB," which is the tare weight of the empty cylinder. Weigh the entire cylinder, subtract the tare weight, and you will know how much propane you actually bought.

Many gauge-like devices are available on the market to display how "full" a propane cylinder is. By now, I hope you recognize why this isn't a very effective mechanism for determining how much liquid propane is left in the cylinder. Somewhat more useful are adhesive liquid crystal thermometers that attach to the side of the cylinder. These will register the level of the colder liquid (once the liquid cools from having boiled off some vapor), providing a reasonable estimate of how full the cylinder is. The most accurate, and difficult, method is to weigh the cylinder subtracting the tare weight. The problem is that most people don't want to leave their propane cylinder sitting on a scale. Intriguingly, the Internet of Things is apparently coming to the rescue with dedicated cylinder weight-sensing devices that send updates to your computer or phone. While the initial set of these devices that have come to market aren't particularly precise, we can expect improvements year over year.

The four most common sizes of portable cylinders in the United States are 20, 30, 40, and 100 lbs (see Table 2-1), with 20 lb cylinders vastly outnumbering the others. (See Figure 2-1.)

TABLE 2-1: Common "portable" cylinders in the United States

	20 LB	30 LB	40 LB	100 LB
Capacity	4.7 gal	7.1 gal	9.4 gal	23.6 gal
Empty weight	18 lbs	24 lbs	29 lbs	68 lbs
Full weight	38 lbs	54 lbs	70 lbs	170 lbs
Height	18"	24"	29"	48"
Diameter	12.5"	12.5"	12.5"	14.5"

FIGURE 2-1: The United States' most common propane container, the 20 lb cylinder

OPD's are required for cylinders produced after September 30, 1998. As of April 1, 2002, the OPD is mandated for refilling a cylinder, so older cylinders without OPDs are not allowed to be refilled. You can tell the difference by looking at the valve handle. If it's triangular, it's a tank with an OPD; if it's a 5-point handle, it's too old to allow refilling.

These consumer cylinders must be requalified 12 years following their manufacture date and every 5, 7, or 12 years thereafter, depending on how the last requalification, if any, was completed.

Cylinders overdue for inspection must not be refilled. In many states, disposal of condemned propane cylinders is illegal. The best way to dispose of a propane cylinder is to have a propane company take it and scrap it.

Safety Note

The OPD closes a valve when an internal float rises; this occurs when the cylinder is 80% full. The OPD will not function properly when the cylinder is on its side. It is not intended to be an operational safety device! It is entirely possible to release liquid propane through the valve of an OPD equipped cylinder. Cylinders designed for delivering propane vapor should **always** be stored upright and never on their side!

FIGURE 2-2: Cylinder collar markings

Propane cylinders have a collar ring to protect the valve, and a foot ring to assure the ability to stand upright. The collar ring also serves as the location of the various markings that are required. (See Figure 2-2.)

An example of the collar markings would be something like you see in Figure 2-2. Table 2-2 shows what each marking means.

TABLE 2-2: Decoding propane cylinder collar markings

DOT	Manufactured to U.S. Department of Transportation specifications
4BA	Cylinder specification (e.g., 4B, 4BA, 4BW and 4E)
240	Cylinder service pressure
M4875	Cylinder serial number
WCW	Manufacturer's name or registered symbol
TW-17.0 LB	Tare weight of empty cylinder (17 lbs)
04-05	Indicates an original manufacture date of April 2005 If there had been an inspector's mark, it would replace the "-"
MUST BE REQUALIFIED WITHIN 12 YRS. OF MFG. DATE	Date for re-qualification Also the area for indication that the cylinder was retested
WC-47.6	Indicates the cylinder will hold 47.6 lbs of water
DT-4.0	Indicates the cylinder has an OP with a 4″ dip tube length

Propane cylinders are all provided with a safety release valve that opens if the pressure gets too high. The 20 lb cylinder in the

Safety Note

Never draw propane from cylinders of the size we're discussing (or larger) without a regulator attached to the cylinder valve. The regulator moderates and reduces the cylinder pressure and serves as a safety barrier between the system and the internal pressure of the cylinder. It is absolutely unsafe to draw propane directly from these cylinders. Commercial products such as roofing tar melting torches and weed burners will skirt this requirement by using a metering orifice before a needle valve. This is not a safe approach for artists or makers! Use a regulator!

United States is specified by the Department of Transportation as having a service rating of 240 psig and a burst pressure of 960 psig (four times the service rating). The safety valve (known as a CG-7) is designed to open between 345 and 465 psig. This is one reason that propane tank explosions are fairly rare. Before the internal pressure can climb to the point of exploding the tank, the relief valve will open and vent the gas.

There are many stories, most apocryphal, about exploding propane cylinders. It is absolutely true that, under extreme conditions, they can explode. However, most people are under misapprehensions about just how extreme those conditions need to be. An exploding propane cylinder would be referred to as a *boiling liquid expanding vapor explosion (BLEVE)*. (See Figure 2-3.)

Propane BLEVEs are exceedingly rare due to the astonishingly tough nature of propane cylinders. Given how common they are, it's a testament to the design and certification of propane cylinders that there are so few accidents. A propane cylinder BLEVE would occur if the pressure inside a cylinder grows so great that it exceeds the ability of the cylinder's safety valve to vent the vapor. The circumstances where a propane cylinder would vent vapor are not uncommon. But vented vapor, even ignited vapor venting at high pressure, are not at all the same thing as a BLEVE.

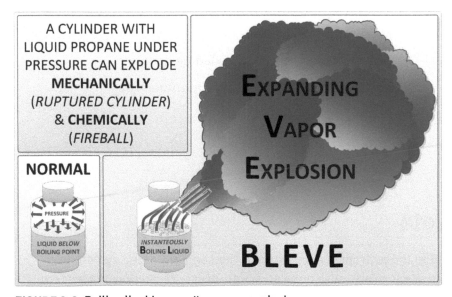

FIGURE 2-3: Boiling liquid expanding vapor explosion

Perhaps one of the strongest testaments to the design of propane cylinders is how rarely propane bobtail trucks have fires or explosions in comparison to the number of times they have accidents. In 2010, some 35,000 bobtails in the United States averaged about 135 rollovers (there were also 7,000 bulk transports and 7,000 cylinder delivery vehicles out there as well). Almost none of these accidents resulted in a gas leak of any kind.

Fire is a definite danger with propane, but explosions aren't what we should be afraid of. We are at much greater risk if we don't follow basic safety procedures that keep unregulated vapor release from occurring.

Hoses, Pipe, and Tubing

Hoses

Propane hoses are also rated according to the pressure they can support. Low-pressure hoses are frequently rated at a maximum of 5 psig. High-pressure hoses frequently have a maximum working rating of 350 psig and must have a 5-to-1 safety factor. According to NFPA 58 National Fuel Gas Code, high-pressure hose will be continuously marked to provide at least

1. LP-GAS HOSE or LPG HOSE

2. Maximum working pressure

3. Manufacturer's name or coded designation

4. Month or quarter and year of manufacture

5. Product identification

(See Figure 2-4.)

Safety Note

The hose must be matched to the appropriate pressure of the segment of the system in which it's operating. Do not use low-pressure hose on a high-pressure section of a system.

FIGURE 2-4: **High-pressure hose markings**

Hoses are typically sold with fittings mounted on each end. It's not safe to use hose clamps or barb fittings with high-pressure propane systems (in excess of 5 psig). Hose assemblies in high-pressure systems must be pressure tested to 700 psig. Special industrial crimping equipment is required to mount fittings safely on high-pressure hose.

Hoses used for acetylene are not usable for propane. Most acetylene hoses are described as grade R hose. Propane requires grade T hose. (Grade T hose will also support acetylene.)

Pipe and Tubing

A variety of piping materials are approved for propane use, though not all types are approved for all uses.

Approved piping materials:

- Wrought iron
- Steel—black or galvanized
- Brass
- Copper
- Aluminum alloy
- Polyethylene (must be marked "gas" and "ASTM D 2513")

Approved tubing materials:

- Steel
- Brass

- Copper—type K or L
- Aluminum alloy (must be above ground and interior use only)
- Polyethylene (must be marked "gas" and "ASTM D 2513")

Cast iron pipe is **not** approved for propane duty.

By far the most common materials used for plumbing propane, at least in residential usage, are black iron pipe and copper tubing. (See Figure 2-5.)

Steel and black iron (aka *wrought iron*) pipe must be of at least standard weight (Schedule 40 per NFPA 54 2.6.2.b) and must conform to one of the following standards:

- ANSI/ASME B36.10, Standard for Welded and Seamless Wrought Steel Pipe
- ASTM A53, Standard Specification for Pipe, Steel, Black and Hot-Dipped, Zinc-Coated Welded and Seamless
- ASTM A106, Standard Specification for Seamless Carbon Steel Pipe for High-Temperature Service

Schedule 80 is a heavier weight pipe with thicker walls. It is sometimes referred to as *extra heavy duty pipe*.

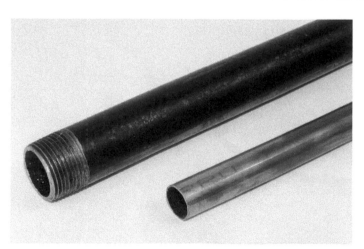

FIGURE 2-5: The most commonly used residential materials: black iron pipe and copper tube

The pressure ratings for piping relate directly to the temperature. Increased temperature reduces the maximum ratings. The size of the pipe and the construction method, material, and type of joints also factor into the maximum pressure ratings. Table 2-3 provides the working pressures under the most common set of conditions. If you are working with pipe or conditions not described below, take the time to investigate the specifics for your circumstances.

TABLE 2-3: Working pressure for Schedule 40 and 80 pipe

CARBON STEEL PIPES—WORKING PRESSURE ACCORDING TO ASME/ANSI B 36.10 AND ASTM A53 B							
Nominal Size (Inches)	Pipe Outside Diameter (Inches)	Schedule Number	Wall Thickness (Inches)	Inside Diameter (Inches)	Working Pressure ASTM A53 B to 400°F		
					Manufacturing Process	Joint Type	psig
½	0.84	40ST	0.109	0.622	Continuous weld	Threaded	214
		80XS	0.147	0.546	Continuous weld	Threaded	753
¾	1.05	40ST	0.113	0.824	Continuous weld	Threaded	217
		80XS	0.154	0.742	Continuous weld	Threaded	681
1	1.315	40ST	0.133	1.049	Continuous weld	Threaded	226
		80XS	0.179	0.957	Continuous weld	Threaded	642
1¼	1.66	40ST	0.14	1.38	Continuous weld	Threaded	229
		80XS	0.191	1.278	Continuous weld	Threaded	594
1½	1.9	40ST	0.145	1.61	Continuous weld	Threaded	231
		80XS	0.2	1.5	Continuous weld	Threaded	576
2	2.375	40ST	0.154	2.067	Continuous weld	Threaded	230
		80XS	0.218	1.939	Continuous weld	Threaded	551
2½	2.875	40ST	0.203	2.469	Continuous weld	Welded	533
		80XS	0.276	2.323	Continuous weld	Welded	835
3	3.5	40ST	0.216	3.068	Continuous weld	Welded	482
		80XS	0.3	2.9	Continuous weld	Welded	767
4	4.5	40ST	0.237	4.026	Continuous weld	Welded	430
		80XS	0.337	3.826	Continuous weld	Welded	695
6	6.625	40ST	0.28	6.065	Electric resistance weld	Welded	696
		80XS	0.432	5.761	Electric resistance weld	Welded	1209

There is a great deal of debate on the Internet regarding the appropriateness of galvanized pipe for propane or gas use. This primarily comes from a concern that the galvanized zinc coating will flake off due to corrosion from the gas and clog valves. This is really a historical concern relating to older gas mixtures and no longer relevant. While I can't speak to every state's or country's regulations, galvanized pipe is clearly allowed under the Liquified Petroleum Gas Code (NFPA 58 5.8.3.1).

The careful reader will note that neither PVC nor PEX is on the list. There are many statements on the Internet stating that PVC can be used for underground gas pipe (often supposedly in Texas). NFPA 54, the National Fuel Gas Code, describes polyethylene as the only plastic fuel supply line that may be used (and only in underground use). The only mention of PVC is in use as regulator vent piping.

Safe use of plastic pipe is beyond the scope of this book. Artists and makers will be more successful in their initial efforts sticking to black iron pipe and copper tubing.

Regulators

Regulators come in a variety of types depending on the needs of the downstream system. The primary factor in determining the appropriate regulator is the pressure required. Regulators will be rated for the pressure they should deliver. While they are often classed as *high pressure* or *low pressure*, it is important to know what actual pressure within those ranges they are delivering. Regulators can deliver a fixed pressure or an adjustable pressure range depending on their design.

Low-pressure regulators typically describe the pressure they deliver in inches of water column (WC, 1″ WC = 0.036127 psig). Common low-pressure BBQ regulators operate at 10.5″ WC (0.3793 psig). (See Figure 2-6.) Quite a step down from a cylinder with an internal pressure that might be at 200 psig!

Some propane systems use a two-stage regulator system. This could be independent first-stage and second-stage regulators (reducing the pressure in two stages), or an integral two-stage regulator that combines both stages in a single frame. Complete systems might include an integral two-stage regulator at the cylinder and a low-pressure regulator downstream to supply a pilot light system. The combinations and designs of regulators are driven by the pressure needs of the system and the various economies of different approaches.

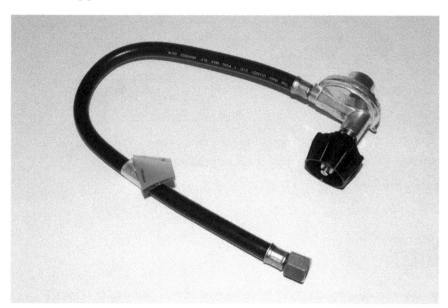

FIGURE 2-6: A common BBQ low-pressure propane regulator

High-pressure regulators are more frequently adjustable than low-pressure regulators. (See Figure 2-7.) Common adjustable high-pressure regulators come in ranges such as 0–10 psig, 0–20 psig, 0–30 psig, 0–60 psig, or even 0–100 psig!

A regulator delivers a consistent pressure even as the cylinder pressure varies. However, high-pressure regulators may be set to deliver a pressure higher than the current cylinder pressure (if the cylinder's propane temperature is low enough). A big, beefy 0–100 psig regulator, wide open at the 100 psig setting, might be delivering a measly 10 psig if you've been drawing so much propane that the temperature has dropped to 20°F (–29°C). Of course, the ice all over the tank will be a big clue to what's going on!

FIGURE 2-7: An adjustable high-pressure propane regulator

Safety Note

Propane regulators are considered damaged if they have been submerged under water. Regulators are not repaired; they are replaced.

Thread Designations

Given that threads are standardized (there are a horrific number of different standards, but by gum, they're all standardized!) you might reasonably imagine that there would be standard names for each type. This is where things get a little confusing. There are specific names that differentiate one standard from another, but there may be many names or acronyms for a particular standard.

For the most part, we will encounter two thread standards in the projects in this book: *National Pipe Thread Taper* and *SAE* flare fittings. Fitting standards around the world vary significantly, so take the time to understand which fittings are available and appropriate if you're outside the United States.

National Pipe Thread Taper

National Pipe Thread Taper (NPT) is an American tapered thread standard (ANSI/ASME B1.20.1). This means that, as opposed to a straight thread, the part of the pipe or fitting that is threaded tapers at a rate of 1 in 16 from base to tip. This causes the joint to pull tightly and create a fluid-tight seal. Common similar international thread standards are British Standard Pipe Taper (BSPT) and Metric Threads, Taper. However, none of these are compatible with each other.

When the acronyms for the male and female versions of NPT threads show up, the confusion starts:

- FPT/MPT
  - Female/male pipe tapered
  - Female/male pipe thread
- FIP/MIP
  - Female/male iron pipe
- FNPT/MNPT
  - Female/male national pipe tapered

Usage is inconsistent even at a single store, sometimes even from a single vendor. The important information is that all of these represent the same thread standard and are compatible with one another.

Straight pipe thread standards are tempting to use with tapered pipe fittings but are **not** intended to work together. The male straight pipe thread cannot penetrate the female tapered fitting far enough to make a good join, and the male tapered fitting doesn't fit tightly into the female straight fitting, causing potential leakage.

I will use *FIP* and *MIP* to designate tapered thread types in this book. My rationale for doing so is completely arbitrary and probably due to some deep-seated irrational urge to use a vowel in my acronyms when possible.

Flare Fitting Threads

The flare fittings used in this book are entirely ⅜″ male and female SAE flare fittings (SAE J512, J513 and ASME B1.1). These are 45° flare fittings and are the specified fittings for use in the United States with propane or natural gas (NFPA 54/ANSI. Z223.1 National Fuel Gas Code).

Internationally there are a number of other flare fitting standards; JIC 37°, JIC 30°, JIS 30°, SAE 37°, 42° Metric Bevel, 45° INF, and more. I'm not able to advise on the locally appropriate replacement for the SAE 45° in locations outside the United States.

Oddly, there don't seem to be common acronyms for male and female flare fittings, even within the United States. In this book, I've used *MFL* and *FFL*. This is probably a travesty and will get me a lifetime of haters. In the immortal words of Kurt Vonnegut, Jr., "So it goes."

Fittings

Metal pipe joints must be threaded, flanged, brazed, or welded. Threads must be taper threads and use appropriate joint compounds (e.g., pipe dope resistant to liquid propane, or yellow gas-rated teflon tape). NFPA 58—the Liquified Petroleum Gas Code—states that cast iron fittings shall not be used. It allows fittings of steel, brass, copper, malleable iron, or ductile iron. Black iron is generally made from malleable and ductile iron; it is sometimes even referred to as *ductile malleable iron (DMI)*. Cast iron, unlike brass or bronze, will not smoothly deform under pressure; it has the potential to develop micro or even visible cracks. The most common fittings that are found in residential systems are black iron, copper, and brass. (See Figure 2-8.)

Soldering is not permitted for propane systems under pressure; only brazing. The brazing material must have a melting point that exceeds 1000°F (538°C) and have a phosphorus content of below 0.05%.

FIGURE 2-8: Black iron, copper, and brass fittings

While brazing is permitted under the NFPA standards, it is exceptionally difficult to tell a well-brazed joint from a poorly brazed joint. Without tests such as hydrostatic pressure testing, many authorities having jurisdiction (AHJs) (such as the fire department or code enforcement departments) will not permit brazed joints. The projects in this book use brazing for its high temperature resistance and only use them for construction of burners that operate at close to atmospheric pressure (14.696 psi or 101325 Pa).

Metallic fittings must comply as shown in Table 2-4.

TABLE 2-4: Allowed Materials

		FITTING						
		Steel	Copper	Brass	Bronze	Malleable Iron	Aluminum Alloy	Cast Iron
Pipe	Steel	Yes		Yes	Yes	Yes		Yes
	Wrought Iron	Yes		Yes	Yes	Yes		Yes
	Copper		Yes	Yes	Yes			
	Brass		Yes	Yes	Yes			
	Aluminum Alloy						Yes	

Other important notes:

- Cast iron bushings are not allowed.
- Cast iron fittings cannot be used in systems containing a flammable gas/air mixture.
- Aluminum alloy fittings cannot be used if threads form the joint.

For the same reasons that I said plastic pipe was out of scope in a previous section, plastic fittings are also out of scope in this one.

Fittings come in a variety of types for a variety of purposes. Threaded, flared, and compression fittings are all common types. (See Figure 2-9.) Technically, flare fittings are a type of compression fitting, but the term *compression fitting* usually refers to a thin wall pipe fitting with a ferrule. Compression fittings are *not* appropriate for gas work in most jurisdictions.

Flare fittings don't require pipe dope or Teflon tape to seal the joint. Flared joints may only be used with nonferrous pipe and tubing (e.g., copper and brass).

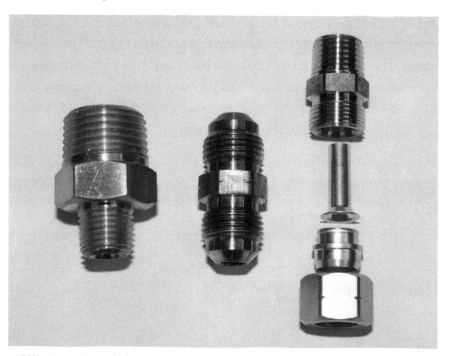

FIGURE 2-9: Threaded, flared, and compression brass fittings

Threaded fittings have code restrictions when used in concealed settings. In this case only, elbows, tees, and couplings are acceptable. Unions, bushings, tubing fittings, swing joints, and "compression couplings made by combinations of fittings" (per National Fuel Gas Code 1211.3.2) are not acceptable.

Additionally, the National Gas Fuel Code states that in concealed or unconcealed settings, all metallic fittings must be used:

- Within the manufacturer's pressure and temperature ratings

- Within the service conditions anticipated with respect to vibration, fatigue, thermal expansion, or contraction

- Installed or braced to prevent separation of the joint by gas pressure or external physical damage

- Acceptable to the authority having jurisdiction (usually your local code inspector or fire department)

Valves

Manual Valves

Valves serve the critical role of isolating segments of a propane system. Code specifications often reference the need to have shut-off valves at numerous points throughout a propane system. This section will cover manual valves; electrically controlled solenoid valves will be covered in a later section. Valves are sold in a variety of dimensions with myriad connection types and a host of pressure ratings. Multiplying varieties × myriads × hosts results in many more valves than could be covered in this text. We will look at the most common valves. As a note, the body of a gas-rated valve is usually brass or bronze. This is because those materials will not spark and will generally deform, rather than crack, under pressure.

Look for valves with the following ratings, usually marked on the valve or handle (see Figure 2-10, next page):

WOG—Water, oil, and gas

CGA-AGA—Canadian Gas Association, American Gas Association

CSA—The newer organization, which is a combination of the CGA and AGA

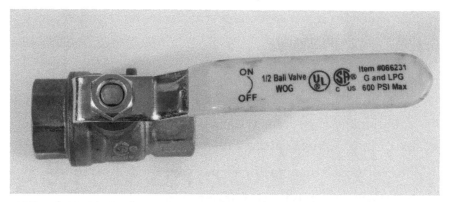

FIGURE 2-10: A ball valve with WOG marked on the handle (it's marked CSA on the side)

Safety Note

When buying valves, make sure that you are confident that the valve is rated for gas at the pressure you will be using it for. Low-pressure systems, meaning systems operating at something less than ½ psi, can use valves with a lower rating, but these must only be downstream of a low-pressure regulator. Don't be cheap with your safety; buy only valves rated for high pressure so you don't accidentally reuse a low-pressure one inappropriately. The valves described below should all be rated at a minimum of 250 psig unless specifically noted.

Cylinder Valves

The first propane valve that most people will encounter is the cylinder valve. All cylinder valves are multi-turn and have a built-in bleeder valve to handle cylinder over-pressure situations. Because the cylinder valve serves to shut off the main supply of propane, it has to meet a number of specifications. However, the standards for cylinder valves have changed over time.

Old-style cylinder valves were described as *POL valves* (named for the originating manufacturer, Prest-O-Lite, aka CGA-510). These valves have internal threads. The POL valve requires a wrench to make a secure union and will vent gas if opened with no fitting attached. The nickname for POL is *put-on-left* since these fittings thread in the opposite direction of standard threads. These valves should have a plug threaded into them during transport. (See Figure 2-11.)

Safety Note

You cannot visually tell whether a cylinder valve is opened or closed. Many accidents have been caused by people attempting to turn a valve, finding it won't move, and therefore assuming that it's closed. For that reason, it is considered a good practice to always open the valve fully, then back it off a quarter turn. This provides a tactile mechanism (the handle turning) to indicate that the system is currently open. In general, people will then verify the closing direction on the handle itself and successfully close the valve.

Newer valves are known as ACME valves (aka *type I outlet*). These valves have external quick closing coupling (QCC) threads and are designed to be able to make a secure attachment by hand. These valves usually have the POL female threading inside for backward compatibility. The newer valves have a back check assembly that will not release gas unless a hose or other fitting is attached.

You can also purchase fittings to attach the small (typically 1 lb in the United States) cylinders to other fittings and plumbing. These are not very commonly used, but have potential for small propane projects. Note that these fittings often have a flow reducer built in-line.

FIGURE 2-11: ACME valve with QCC, POL valve, portable cylinder connector

Quick Connect Valves

FIGURE 2-12: Low- (left) and high- (right) pressure gas-rated quick connect fittings

The type II outlet is a propane-rated quick connect coupling. (See Figure 2-12.) It allows a hose, or other component attached to the male fitting, to be easily decoupled from a pressurized source without gas leaking. Propane-rated quick connects are fantastic for quick setup and teardown of systems. They are considerably more expensive than quick connects used in compressed air systems. Compressed air quick connects are not safe for propane use due to their frequent use of rubber seals.

Be careful; gas-rated quick connect fittings are also pressure-rated. A low-pressure quick connect is typically only rated to ½ PSI and will blow its seals (and waste your money) in a heartbeat if used in a high-pressure system.

Safety Note

Did I mention not to be cheap with your safety! Don't use quick connects intended for compressed air in your propane system. They don't have the appropriate pressure ratings and usually have seals that can corrode in contact with LPG.

Ball Valves

FIGURE 2-13: Cutaway view of a ball valve
Archivo fotográfico Standard Hidráulica. Free of copyright.

The most common valve used to isolate segments of a system is a quarter-turn ball valve. These valves typically have a straight handle that's parallel to the hose or pipe when open and perpendicular when closed. Ball valves have an internal . . . wait for it . . . ball, with a hole drilled through it. (See Figure 2-13.) When turned, the hole drilled through the ball is either in line with the passage or 90° away from it. Ball valves are extremely durable and are able to completely shut off gas flow with a single, quick motion. They also give a fast visual confirmation of their state. When the handle is in line with the body, the valve is open; when perpendicular, it's closed. (See Figure 2-14.) For these reasons, quarter-turn ball valves are the preferred (and in some cases mandated) valve used as a manual safety in propane systems.

It is tempting to try to use a partially open ball valve as a means to regulate flow. This is a misuse of the ball valve; it should only be used in fully open or fully closed mode.

Needle Valves

Needle valves are designed to regulate flow. (See Figure 2-15.) While inappropriate for direct connection to a cylinder (use a regulator!), needle valves of the appropriate pressure rating can serve to vary the pressure downstream from the regulator. Needle valves differ from regulators in that a regulator should maintain the output pressure consistently as the supply pressure varies (until the supply pressure drops below the regulated pressure). A needle valve provides a proportional reduction in flow and the output will vary with the supply pressure. Since the needle valve should be downstream of a regulator, this shouldn't be a big problem.

FIGURE 2-14: **Open ball valve**

FIGURE 2-15: **Common needle valve**

Safety Note

Some propane tools, such as asphalt-melting torches, appear to have a needle valve as the only means of regulating the pressure from the cylinder. This is because the manufacturer has built both a tiny metering orifice and a needle valve into the device and tested it extensively to protect the company against liability claims. Don't take the chance; use a regulator.

Needle valves are typically less expensive and have been around a lot longer than regulators (see Figure 2-16, next page), so they are excellent for placing close to a torch, pilot, or other output when you'd like to fine-tune the pressure.

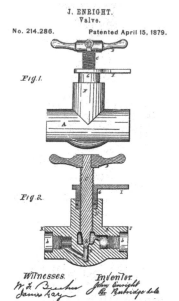

FIGURE 2-16: Needle valve patent image from 1879

FIGURE 2-17: A gate valve

Gate, Bibb, and Other Water Valves

Gate valves are multi-turn valves most commonly used on exterior water spigots. They can also be used to control water throughout a plumbing system. They're frequently found under sinks and on hot water heaters. They are sometimes referred to as bibb valves since they often have a hose bibb as a fitting so that you can connect a garden hose. (See Figure 2-17.)

The temptation to use water valves is tremendous. They are plentiful and cheap. But they aren't intended for gas and immediately demonstrate to any inspector that your device is constructed with unrated parts.

Safety Note

This is worth reiterating. The valves commonly found at hardware stores for use with water supplies are not intended for use in propane systems. While rare variations are propane-rated, you would have to make a considerable effort to find them. Don't succumb to temptation and use valves intended for water in your propane system. They aren't appropriately pressure-rated and may well have rubber seals that propane can corrode. Your local hardware store should have gas-rated ball valves; use them!

Solenoid Valves

Electrically activated valves are typically driven by solenoids. A solenoid is an electromagnet that moves a rod that opens and closes a valve. Solenoid valves come in a variety of configurations, including:

- Alternating or direct current
- Voltage (ie, 12 VDC, 120 VAC, 24 VDC)
- Whether the valve is normally open (NO) or normally closed (NC) when no power is applied

- Pipe diameter (e.g., ¼", 1", 10 mm)

- External fittings

- Seals (aka *diaphragm*)

- Direct acting, internally piloted, or manual reset

- Two-way, three-way, manifold—basically, how many holes it has. The two-way has one in and one out. The three-way has one in and two out, and so on.

- Temperature range

- Enclosure type

- Rated pressure

For general use in basic propane systems, the two-way, direct acting, 12V DC, normally closed, gas-rated solenoid is the work-horse of choice. (See Figure 2-18.) While a number of different seals are gas-rated, nitrile rubber (aka, *NBR* and *Buna-N*) is commonly used in propane applications. The pipe diameter would be chosen for the task at hand and the fittings would be chosen to match the rest of the system.

FIGURE 2-18: **12 VDC gas-rated solenoid valve**

Ideally, a solenoid used for propane control would have a pressure rating greater than 250 psig (265 psi). However, NFPA 58 states that as long as the usage is restricted to propane vapor at 125 psig (140 psi) or less, the minimum service pressure rating for valves, fittings, pipe, and tubing is 125 psig (140 psi; NFPA 58, table 5.8.4.1.) 130 psig (145 psi) is a common rating for solenoids since that's pretty close to 10 kilograms per square centimeter, or 1 mega pascal. Using solenoids rated at these levels significantly reduces the cost without compromising safety (as long as your system stays below the rated pressure).

Solenoid valves provide an interface between the mechanical and electrical domains. Being able to open or close a valve with electricity means that the valve can be manipulated from a distance or controlled by a computer. These valves can also be opened and closed faster than most mechanical valves, allowing for potentially fine-grained control.

I could have referred to this section as "Electrical Valves" and included additional discussion of proportional valves. A solenoid valve is usually a basic on/off valve. Other types of electrically controlled valves offer variable control. These valves are generally very expensive, fairly slow, and difficult to obtain. For those reasons, I am not including them in this text.

Accumulators

NFPA 160, *Standard for the Use of Flame Effects Before an Audience*, defines an *accumulator* as:

> *3.3.1 Accumulator. A container or piping that holds a predetermined volume of fuel that is ready for use in a flame effect.*

(In the context of this book, accumulators store propane vapor.)

To understand the purpose of an accumulator, think about how much vapor you can access in a system. The source of the propane vapor we start with is the propane cylinder. However, as noted earlier, these cylinders contain both propane liquid and vapor. If your system needs to release more vapor than the cylinder contains, you have to wait until it boils off the liquid.

Accumulators provide one or more additional locations to store vapor. This is usually for the purpose of achieving a bigger fireball in something like a flame effect.

Accumulators store vapor at the pressure of the section of the system they are attached to. Saying that another way, if your system consists of: cylinder (at 140 psig) → regulator (set to 50 psig) → accumulator → solenoid valve → vent, the pressure in the accumulator is 50 psig. (See Figure 2-19.) Pretty simple, but important. You should always maintain an awareness of what the pressure should be (and is) within the various segments of your system.

NFPA 160 defines a number of important considerations for using accumulators safely in section 9.3.2.6, *Systems Using Fuel Accumulators*.

This section states that accumulator tanks must have been built to be pressure vessels and have been tested under ASME or DOT standards.

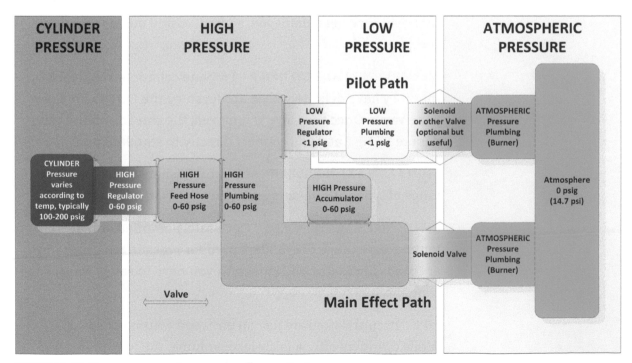

FIGURE 2-19: Pressure zones in a propane flame effect

The standard also states that the accumulator can't be bigger than the amount of fuel needed for the effect. Manual shutoff valves are also required on the accumulator.

Accumulators have to be constructed so that they don't allow any air (i.e.,oxygen) to enter and create a flammable mix. Operators aren't allowed to fill the accumulator until immediately prior to firing the effect. A method to safely discharge the accumulator is also required.

Unlike the other components described in this section, accumulators aren't something you typically buy on eBay or at Home Depot. So what constitutes a safe accumulator?

DOT-APPROVED CYLINDERS DOT-approved cylinders are marked. There will be a cylinder type such as DOT-3A, DOT-3A480X, DOT-3AA, DOT-3B, DOT-4A, DOT-4AA480, DOT-4B, DOT-4BA, or DOT-4BW, followed by a number representing the service pressure rating (which is espressed as psig). Make sure you select a service rating over 250 psig. It is also important not to trust cylinders that may have been damaged, abused, rusted out, or otherwise compromised.

AN EMPTY PROPANE CYLINDER Propane cylinders are ideal in that they have definitely been approved by the DOT to hold propane vapor. Once the valve is removed, the threads on the cylinder are ¾″ NPT and will connect directly to standard pipe.

EMPTY FIRE EXTINGUISHER Fire extinguishers come in a wide range of types of cylinders and many are DOT approved.

INDUSTRIAL GAS CYLINDERS Once safely emptied of their contents, a wide range of cylinders used for pressurized gas storage are good candidates. Did I mention you have to safely remove the contents?

PIPE Accumulators are just an enclosed volume of space. While cylinders offer a fairly large volume, the very piping you use in your system serves as a volume that stores vapor. Adding additional pipe beyond the minimum necessary, or using a larger diameter pipe than needed, will increase the amount of vapor stored for release. The pipe (along with the rest of your system) has to have appropriate pressure seals.

Safety Note

To use DOT cylinders as accumulators, an existing valve on the cylinder is usually removed. You **MUST** be ABSOLUTELY SURE that the cylinder is empty of any dangerous or combustible contents before attempting to remove a protective valve. The best way to safely remove the valve is to have a professional dealer remove it. **DO NOT** attempt valve removal without fully understanding the safe methods of content removal.

Reusing Propane Cylinders as Accumulators

The most common accumulator used in DIY flame effects is a repurposed propane cylinder. (See Figure 2-20.) However, there is an astonishing amount of bad advice on the Internet about reusing propane cylinders for various purposes. In our goal to dispel falsehoods and understand things from first principles, let's break down the most commonly recommended methods for acquiring an empty cylinder according to how safe or risky they are.

FIGURE 2-20: Repurposed propane cylinder used as an accumulator

The Safe Methods

GET EMPTY CYLINDERS FROM A PROPANE DEALER.   This is ideal and absolutely my recommended method. A purged cylinder with the valve already removed is every goal fulfilled in the safest manner possible. Most dealers will sell decommissioned cylinders for scrap prices. I usually purchase them for $5 each. Compare any risk you might want to take to acquire an empty cylinder with spending $5. Go buy an empty cylinder from a dealer if you need one.

BUY A BRAND-NEW EMPTY CYLINDER. This is the next best option. You're guaranteed a good cylinder and totally sidestep the issues of flammability. You still have to remove the cylinder valve, but this is a matter of securing the cylinder (with a ratchet strap or a pipe through the handle) and unthreading the very stiff valve. This is best accomplished with a special wrench that fits down over the top of the valve and allows a bar to turn with leverage above the height of the collar guard.

PURGE THE CYLINDER, THROUGH THE CYLINDER VALVE, REPEATEDLY WITH AN INERT GAS. Nitrogen is the gas of choice but CO_2 is a source easier for most artists and makers to acquire. Three or more purges at 40 psi will bring the concentration of propane vapor down to the point where, even if the cylinder is then left open to the air, the propane will be below the limits of flammability.

I want to discuss the other methods that come up on the Internet. None of them are remotely worth the risk. The bottom line is: **DO NOT TRY THESE METHODS**. Any of them could result in tens or hundreds of thousands of dollars in medical bills or liability settlements. Many people may argue, with varying degrees of validity, that these methods are appropriate. I am not willing to do so in a text offered as an introduction to propane.

Safety Note

An "empty" cylinder with only combustible vapor is, in many ways, more dangerous than a cylinder with liquid and vapor. A cylinder with liquid will continue to replenish the vapor and maintain a percentage of vapor above the flammability point. As we discussed in Chapter 1, "Understanding Propane," once vapor mixes with sufficient air, it becomes flammable. An "empty" cylinder—that is, one without liquid that still has vapor in it—is safe as long as the cylinder contains only propane vapor. As soon as the valve is removed, air can enter and mix with that vapor. When that mix becomes flammable, the cylinder is becomes potentially explosive and is extremely dangerous.

If you encounter someone with a compelling argument about the "safety" of these methods and decide to pursue them, please investigate carefully and estimate the risks rationally. My advice remains; *don't try them.* YouTube doesn't show the part where you're in the ambulance, hospital, or rehab.

Bad Ideas (in Decreasing Order of Danger)

REDUCE VAPOR PRESSURE TO ATMOSPHERIC BY OPENING THE VALVE. This is a common suggestion on the Internet. Many suggest letting the cylinder sit for some period (ranging from days to years). With the old-style valve, you could just leave it open. With the newer valves, you would have to thread a fitting into the cylinder valve to get it to remain open. Remember that propane is heavier than air, so the vessel would need to be upside down to drain the remaining vapor. During an indeterminate period while the vapor was hopefully draining, it would be unattended in a flammable state. This is risky and potentially irresponsible. If you haven't drained all the propane, you'll be attempting to unthread an astonishingly difficult valve that has "liquid weld" holding it place. Any sparks from slipped wrenches or other sources could ignite the gas in the tank or in your face. Take the cylinder to your propane dealer; most of them will swap it for a decommissioned cylinder.

PURGE THE CYLINDER REPEATEDLY WITH COMPRESSED AIR. Special fittings would be required, but ideally this would be considerably quicker than just opening the valve and waiting. However, you still have to cross the risky valley of flammability. Mixed with air, even a small amount of propane vapor remaining in a 20 lb cylinder has shocking energy content. If only 10% of the propane vapor remained after all the pressure was released, 183 BTU of energy is sitting in a very stout vessel for some period of time (as long as it takes to pressurize the vessel beyond the flammability mix, which is from 0–65 psig in the fastest method I could devise). 183 BTU is pretty close to the energy content of 64 g (¼ cup) of black powder. Don't be a cheapskate; spend the money and buy an empty cylinder.

Reviewing Commonly Misused Parts

Hopefully, the difference between safe and unsafe parts is becoming clear. It's worth a quick review to look at the most common unsafe parts that get used in amateur devices.

FIGURE 2-21: White versus yellow Teflon tape

WHITE TEFLON TAPE (SEE FIGURE 2-21)	
Why Do People Use It?	Cheap and easy to find in almost any hardware store or section.
Why Shouldn't You Use It?	It shreds. White tape is the same material as the (appropriate) yellow tape, but it's much thinner. The ends tend to shred into tiny streamers when it's torn or threaded. These tiny Teflon streamers enter and clog the valves and fittings. Ruining a $60 regulator because of white tape is one of the many sad and dangerous potential outcomes.
What Should You Use Instead?	Yellow gas-rated Teflon tape.

FIGURE 2-22: Cast iron vs. brass bushings

CAST IRON BUSHINGS (SEE FIGURE 2-22)	
Why Do People Use It?	Easy to find, common in hardware stores, and convenient to use in an awkward situation when changing from one diameter to another.
Why Shouldn't You Use It?	Because bushings are sandwiched between an external fitting and an internal pipe (or fitting), they are subject to incredible force when torqued. Cast iron is a brittle material. Under torque, cast iron will develop cracks. Under pressure, these cracks can either leak or fail catastrophically. Since the bushing is hidden by the external fitting, inspection is exceptionally difficult.
What Should You Use Instead?	Reducing bells are the appropriate item to use when changing dimensions. In some specific situations, a brass bushing can be used, since brass is soft enough to deform rather than crack.

FIGURE 2-23: Soldered vs. brazed fitting

SOLDERED FITTINGS (SEE FIGURE 2-23)	
Why Do People Use It?	Soldering, or "sweating," copper for water plumbing is a common and easy practice.
Why Shouldn't You Use It?	Soldered fittings aren't rated to hold gas pressures. Additionally, depending on where the fitting is in the device, the melting point of the solder may be exceeded.
What Should You Use Instead?	Brazed fittings with a melting point in excess of 1000°F (538°C) and a phosphorus content below 0.05%. Brazing is a fantastically useful and remarkably easy skill to learn.

HOSE CLAMPS (SEE FIGURE 2-24)

Why Do People Use It?	Convenient, easy to acquire, appears to be cheaper than buying hose with fittings.
Why Shouldn't You Use It?	Hose clamps aren't rated for gas pressures. Homemade hoses are a tremendously dangerous potential point of failure.
What Should You Use Instead?	If you're trying to work with unfitted hose, the first consideration is that you really just shouldn't. Use rated hose with appropriate fittings. It is technically possible to buy unfitted hose and make custom sections, but it requires expensive swaging equipment to mount fittings correctly. Unless you're making a lot of custom hose runs, it's cheaper and safer to buy pre-fitted gas hose.

FIGURE 2-24: Hose clamps

PLASTIC VALVES OR SOLENOIDS (SEE FIGURE 2-25)

Why Do People Use It?	Cheap; easy to buy manual plastic valves at the hardware store; easy to scavenge plastic solenoids from washing machines and other devices.
Why Shouldn't You Use It?	Plastic valves are typically designed for water or other liquids at pressures far below gas ratings. Plastics may also be of a type that is corroded by propane.
What Should You Use Instead?	Metal (typically brass) solenoids and valves with fittings or diaphragms that are designated as rated for gas. Seals made of Buna-N (aka NBR, nitrile rubber) can withstand propane, whereas rubber seals will degrade and corrode.

FIGURE 2-25: Plastic versus brass valves

COMPRESSED AIR PARTS (SEE FIGURE 2-26)

Why Do People Use It?	Easy to acquire, cheap.
Why Shouldn't You Use It?	Fittings designed to be used for compressed air are typically unrated for gas pressures. They frequently also have rubber seals that will corrode under contact with LPG. Quick connect fittings for air compressors are particularly tempting (given the price of gas-rated quick connects), but are not worth the safety risk.
What Should You Use Instead?	Gas-rated fittings with appropriate seals.

FIGURE 2-26: Compressed air (left) versus gas-rated (right) quick connect

FIGURE 2-27: Compression fittings

COMPRESSION FITTINGS (SEE FIGURE 2-27)	
Why Do People Use It?	Easy to acquire.
Why Shouldn't You Use It?	Compression fittings that use an internal sleeve are generally unrated for gas pressure use. Technically, NFPA 54, the National Fuel Gas Code 5.6.8.4 (8), states that compression fittings that are "acceptable to the Authority having jurisdiction" and that are "used within the manufacturer's pressure-temperature recommendations" are acceptable. However, these are typically not the compression fittings that are found in most hardware or plumbing stores.
What Should You Use Instead?	Flanged fittings. Flange tools and flanged fittings appropriate for gas use are available at most hardware and plumbing stores. With a little practice, most makers or artists can make and use safe gas-rated flanged fittings for copper pipe.

DIY Propane Tools

Acquiring tools is one of life's true joys—an endless quest filled with thrills, heartache, adventure, and remorse. There are many schools of thought about tools, and finding the approach that works for you is important. Many people will only use tools of the highest quality, treating them with the utmost care. Other people can be haphazard with their tools, abusing them in various ways. I must confess that I have swung both ways.

I have tools that I treat like holy relics, and imported discount tools that have been involved in unspeakable acts. The last couple decades have seen an explosion of budget tool stores selling products that draw the ire of many purists. But here's the truth: I buy tons of stuff from them. I invest in high-quality tools that I know I'm going to use a lot or need professionally. But acquiring a tool that I'm only going to use a couple times at a fraction of the price of the high-end version is too much for me to pass up. I've also had some awesome fun buying something super cheap and modding it or tweaking it into a great tool.

For what it's worth, I'm absolving you of any shame purists will throw at you if you buy budget import tools. I respect the need for professionals to invest, but makers gotta make and tools are enablement. Go for it!

The following list describes the key tools commonly used when working on DIY propane projects. We'll also look at specific tools needed for various projects in their respective chapters. Keep in mind that there are any number of valid substitutions (i.e., other tools that will work just fine) if you're conscious of what you're trying to accomplish. It's perfectly okay if you don't run out and buy every one of these tools (just because I did). Building projects with well-tooled friends or at your local makerspace can be a great way to get access to skills, knowledge, tools, and people.

TWO CRESCENT WRENCHES (See Figure 2-28.) There will be many times when tightening or loosening two fittings that you will need two crescent wrenches. While an open-end wrench could serve this purpose, it's usually cheaper to start off with two crescent wrenches. Preferably, buy at least one wrench with more than a 1 ¼" (30 mm) jaw opening. This can be useful when using large fittings like pipe unions.

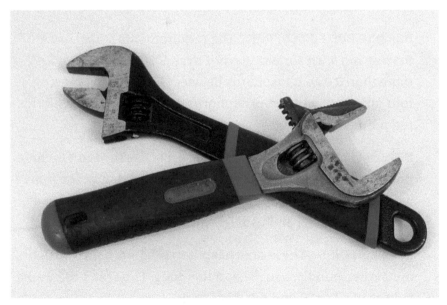

FIGURE 2-28: Two crescent wrenches

FIGURE 2-29: Two pipe wrenches

FIGURE 2-30: Tubing cutter

TWO PIPE WRENCHES   (See Figure 2-29.) Pipe wrenches grip the pipe itself, either turning it or holding it while a fitting is turned. Typically, pipe wrenches are used on black iron or galvanized heavy wall pipe. Copper tubing is generally too thin for pipe wrenches and is rarely threaded. I like having at least one big pipe wrench and one small one.

TUBING CUTTER   (See Figure 2-30.) This tool is used for cutting copper tube. There are some great cutters on the market today that are for one-handed use but only fit one size of tubing. I'm going to recommend that you at least start with the old-school tubing cutter. It will fit a wide range of sizes and usually includes a basic deburring tool.

A BRAZING KIT   (See Figure 2-31.) Consisting of:

MAP-PRO TORCH   Brazing for use with propane requires a melting point in excess of 1000°F (538°C). To easily reach this temperature you really need a hotter burning gas than propane. The most accessible of these is Map-Pro. You'll be happier if you buy a torch with a trigger ignition. A swirl flame head does a better job of heating things up than a straight-through head.

BRAZING ROD FOR COPPER   The requirements state that a brazing rod with a melting point over 1000°F (538°C) and no more than 0.05% phosphorus be used. Silver-based brazing rods (aka BAg) without phosphorus are the best bet to fill this requirement.

BRAZING FLUX   Since the rods we're discussing don't include phosphorus, which acts as a flux, we'll need to flux the joint with some flux compound appropriate for brazing. These are often borax-based and usually water soluble.

FLUX BRUSHES   These are cheap horse-hair brushes with a metal tube handle, usually about 6″ long. They tend to be sold in bags of 10 or 20 and are disposable.

FIRE BRICK   Fire bricks provide a safe base for high-heat operations like brazing. While brittle compared to a standard brick, fire bricks are very useful and can still be useful even if they break. Surprisingly, effective forges and foundries can be built using not much more than stacks of fire bricks.

SAFETY GLASSES (APPROPRIATE FOR BRAZING) Just don't do things in the shop without safety glasses. In this case, there is the added requirement of protecting our precious eyes while torch brazing. OSHA requires a #3 shaded lens. You can use

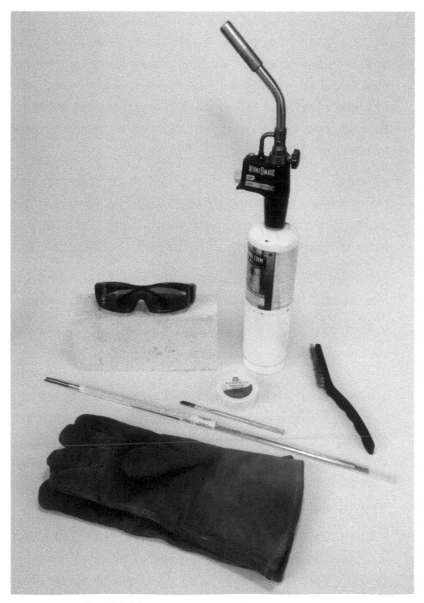

FIGURE 2-31: Brazing kit

a higher shade, but the higher the shade number, the darker it gets. For torch brazing, #3 is the minimum listed shade. Sunglasses are not the same thing. Brazing safety glasses protect against both splatter and the infrared components of the spectrum that can harm the eye. I own a lot of different safety glasses to handle my different needs: welding, plasma-cutting, laser-etching, brazing, grinding, drywall splatter—the list goes on. Don't be cheap when it comes to protecting your sight.

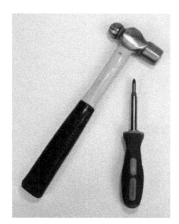

FIGURE 2-32: **Hammer and screwdriver**

GLOVES   Heavy leather gloves are essential when working with high temperatures. Welding gloves are by far the best option and are usually available for budget prices at discount import tool stores. Keep in mind that you can get burned through a welding glove if you hold a part heated to brazing temperature for too long.

HAMMER   (See Figure 2-32.) Pretty much any hammer will do. I once wrote a blues song called, "I Got a Ball Peen Hammer on a Claw Hammer Job," so I'm showing a picture of a ball peen hammer (plus it's fun to say, try it; "ball peen hammer").

SCREWDRIVER   Nothing special, just a useful screwdriver. I like the kind that has four interchangeable bits included.

FIGURE 2-33: **Vise**

VISE   (See Figure 2-33.) A reasonable vise is incredibly useful. Vises range in size from tiny to monstrous. Personally, I want all of them, but a good recommendation for a first vise is a 2½" (60 mm) table swivel vise. This type of vise mounts to surfaces with a built-in clamp instead of being permanently installed. Don't get me wrong; my 5" (130 mm) bench vise gets more use and abuse than any tool in my shop, but I have two small clamping table vises, as well, and couldn't do without them.

FIGURE 2-34: **Full face shield**

FULL FACE SHIELD   (See Figure 2-34.) Eye protection is critical, but I'm kind of partial to the rest of my face, as well. Over the years, I've started using a full face shield for almost everything. For a variety of reasons they're more comfortable than clear safety goggles (for situations where a tinted safety glasses are needed, wear them under the shield).

DEBURRING REAMER (See Figure 2-35.) While the tubing cutter listed above has a knife blade deburring tool, the truth is that it's a massive pain to use. A handheld, conical insert-style deburring tool is much easier to use. It will work on the internal or external edge of the cut, and the results are way better than trying it with the knife blade on the pipe cutter.

FIGURE 2-35: Deburring reamer

I hope that this review of propane equipment and parts has been helpful. New products are constantly entering the market and old favorites often seem to disappear. Keep in mind the basic safety principles and you should be able to judge whether or not something is appropriate. You are likely to build up a collection of parts and tools as the years go by, so always inspect your gear to make sure it's in good operating condition before use.

3

The Low-Pressure Source

THIS BOOK PRESENTS A series of projects that, for the most part, attempt to enhance your skills in an incremental and progressive manner. To accomplish this, the first set of projects uses a low-pressure regulator on the propane cylinder as our fuel source. This will provide propane vapor to our projects at approximately ½ psi.

After we've built low-pressure projects that introduce the concept of how to correctly assemble gas-rated fittings and perform skills like brazing, we'll move to high-pressure projects. In both cases, all the projects in a section will use the same pressure source. This allows us to reuse the same module without having to rebuild a new one for each project. The low-pressure source, which we will describe below, swaps the low-pressure regulator for an adjustable high-pressure regulator to upgrade to the high-pressure source. We'll then use the low-pressure regulator as a component of the pilot system in the boosh flame effect project.

Working with low pressure makes things easier when you're just starting out. However, there are still important safety concerns, regardless of the pressure provided by the regulator. No matter what type of regulator you attach to it, all propane cylinders at the same temperature have the same pressure (see Chapter 1, "Understanding Propane"). So you're working with cylinder pressure in all projects. Additionally, low-pressure regulators still produce a significant amount of gas. Imagine a normal propane outdoor grill; that's a low-pressure system and it puts out considerable gas and heat. All the normal cautions related to working with propane are required for low- as well as high-pressure systems. Nevertheless, it helps to build confidence at ½ psi before we move up to 60+ psi.

You will note that I am specifying the use of hose rated for high-pressure use even though this is a low-pressure source. I'm doing that for two reasons. The first is that we will reuse the hose when we upgrade this to a high-pressure source. The second is that I believe it's dangerous to have a mix of high- and low-pressure hoses in your propane gear. The high-pressure hose works fine for low pressure but the reverse is most definitely not true. It's better to invest a little more and know that your hoses work for all situations.

As much as possible, I've tried to stick to using parts available at local, big box building supply stores. Unfortunately, in the last few years these stores have drastically cut the variety of items that they carry. At the same time, many local plumbing supply stores have closed their doors. This is very sad; I've spent untold hours haunting the shelves of old-school plumbing supply stores to find treasures hidden on dusty shelves. The good news is that Internet vendors have stepped in with a range of products, and a set of prices, that the local stores could never offer. I recommend spending whatever you can afford locally to try to keep good stores alive. But I have no doubt that you'll end up buying a larger and larger portion of your fittings and valves online. (See the appendix for a list of online vendors I've had success with.) Generally, no single online vendor will always have the best price on every item. Always factor in the shipping costs before deciding which vendor to go with.

This diagram describes the major components using industry standard symbols. A schematic is typically included in a burn plan for a flame effect. The low-pressure source has a very simple schematic (see Figure 3-1).

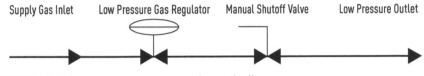

FIGURE 3-1: Low-pressure source schematic diagram

Parts

The low-pressure source has a small number of components.

Low-pressure source parts:

REF	ITEM	QTY
F1	Brass adapter ⅜″ MIP × ⅜″ FIP	1
F2	Brass bushing ⅜″ MIP × ¼″ FIP	1
G1	Propane cylinder	1
G2	10′ high-pressure propane hose ¼″ MIP × ⅜″ FFL	1
R1	Low-pressure propane regulator	1
V1	Gas-rated ball valve ⅜″ FIP × ⅜″ FIP	1

Throughout this book, I'll present a schematic and block diagram for every project. The schematic conveys the essential function; the block diagram the specific parts. (See Figure 3-2.) You can also use the block diagram as an assembly guide.

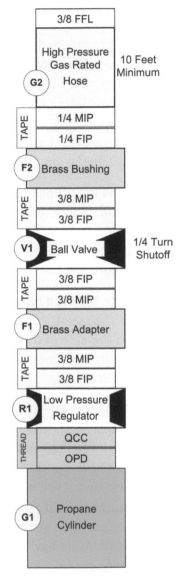

FIGURE 3-2: Low-pressure source block diagram

Taping and Tightening Fittings

TAPING JOINTS

As I mentioned earlier, I'm a taper, not a doper. Therefore, most of the joints in this book, other than the flare fittings, rely on yellow Teflon tape to become gas-tight. Taping a joint correctly is easy to do. Four wraps of tape clockwise around the fittings will do the job. (See Figure 3-3.)

Hold the fitting to be taped in your left hand. (Sorry, lefties, please use your enhanced creative skills to mentally flip this so you're looking at the threads.) (See Figure 3-4.)

With the tape hanging off the back side of the spool, put the edge of the tape on the side of the fitting facing you. Hold the corner with your thumb. (See Figure 3-5.)

Wrap the tape around the threads four times. Wrap with enough tension to allow the threads to make a sharp crease (but don't overdo it). The tape may slip on the first turn if you pull too hard; use your thumb to hold down the first wrap until you cover up enough for the tape to hold itself. Avoid having tape hang over the inner passageway of the fitting. (See Figure 3-6.)

When the wrapping is completed, pull the tape until it breaks itself off at the back of the threads.

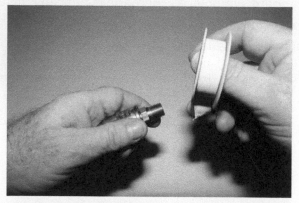

FIGURE 3-3: Taping, step one

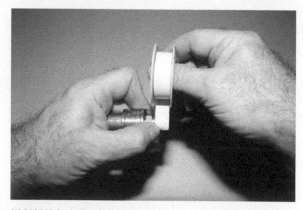

FIGURE 3-4: Taping, step two

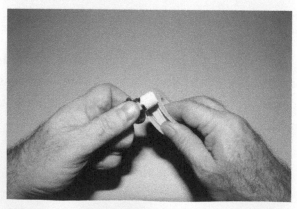

FIGURE 3-5: Taping, step three

If you take apart a taped joint, use a wire brush to get all of the old tape out of the threads (on both the male and female sides of the joint). **Never retape over old tape.** Tape is relatively cheap. Pull a joint apart and retape if you need to; it's better to use a little extra tape than to have an unsafe joint.

TIGHTENING

One question that comes up a lot is, "How tight do I tighten the fittings?" Unfortunately, the answer is, "Tight enough to stop

FIGURE 3-6: Taping, step four

the gas from leaking." This isn't typically described in terms of *torque*, the normal method of expressing how hard to turn a nut or bolt, and therefore how tight to make something. It's generally something that you develop a feel for by tightening and testing for leaks. The feel of "tight enough" is both distinct and difficult to express in words. I typically tighten a taped NPT fitting until it starts to feel like it won't turn much more, and then I give it a turn or two more. I'm vague because the difficulty varies tremendously between fittings and people's ideas of *difficult*. What feels difficult or tight will be different for a tall woman who lifts weights versus an out-of-shape guy who watches too much anime and types on a computer all day (ouch, this is hitting too close to home). Tighten it and then leak-test under pressure to get a feel for it.

Flare fittings are different than taped fittings. Tighten a flare fitting hand-tight, and then just a small amount more until it feels snug. This is usually less than a quarter turn. Technically a ⅜" flare fitting requires 15–25 lbs/foot (20.4–34 Nm) of torque.

Frequently, there's another factor to consider. Do the controls such as ball and needle valve handles line up the way you want them to? The difference between fittings all lined up with the handles oriented in the same direction and a bunch of valves randomly rotated at different angles is dramatic. The orderly stack looks significantly more professional. It can also be important when you consider ease of access to critical controls on a lit system.

The important tip here is to always use a wrench to brace a part you don't want to have move. Use the second wrench to tighten the next part into it. If you have a long sequence of fittings, it's usually best to tighten each one individually, but if you're trying to line things up, it's surprising how often you can tighten at each end of a sequence and eke out an additional quarter turn or more.

Construction

We will assemble the pressure source before attaching it to the cylinder.

1. Start by taping and threading both ends of the male-male brass adapter ⅜" (F1). Tighten one end into the outlet of the low pressure regulator (R1) and the other end into the ball valve (V1).

2. Tape the brass bushing (F2) with yellow Teflon tape and thread it into the ball valve (V1).

3. Tape the male threaded end of the hose (G2) and thread it into the bushing on the end of the ball valve.

That completes the construction of the low-pressure source. (See Figure 3-7.)

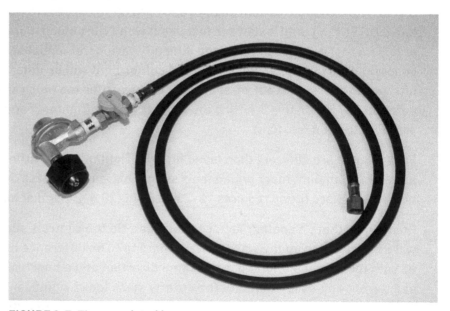

FIGURE 3-7: The completed low-pressure source

Building a Leak-Testing Kit

Any clean empty spray bottle will suffice as your leak-testing kit if you fill it with a mixture of water and dish soap. The ratio of water to soap is not critical. About ½ teaspoon (2.5 ml) per cup (235 ml) works well. A little more or less is fine. Mix this into the spray bottle and shake well. Put the spray head on mist. This, and a towel to wipe up afterward, completes your leak-testing kit. (See Figure 3-8.)

You can buy premixed leak-detecting fluid; it works great but it's hard to justify the cost. You can buy electronic gas detectors, but they don't work well in wind and they're tedious to use when you need to determine a specific joint to tighten. All in all, your homemade spray bottle, soap-and-water test kit will perform as well or better than commercial solutions.

FIGURE 3-8: The leak testing kit

Testing

1. Put your safety glasses on. Verify that the ball valve is turned off.

2. Attach the regulator's QCC fitting to the propane cylinder without Teflon tape. If you've purchased a regulator with a large plastic grip around the fitting, it is threaded by hand. If your regulator does not have the hand grip, tighten with a wrench, but do not overtighten. Some regulators end up not getting propane through the fitting when it is overtightened.

3. Open the cylinder valve all the way and then back off about a half turn. This serves to avoid a rare but possible situation where the valve, tightly up against its threads, freezes locked in the open position and also provides a manual way to determine that the valve is open.

4. Using the leak-test kit, spray the fittings from the cylinder to the ball valve. (See Figure 3-9.) Check for bubbles. If you see any, depressurize and tighten the joint until they stop. A valuable addition is to add a ⅜" flare plug to cap the end of the hose; this will allow you to test the ball valve-to-hose connections, as well.

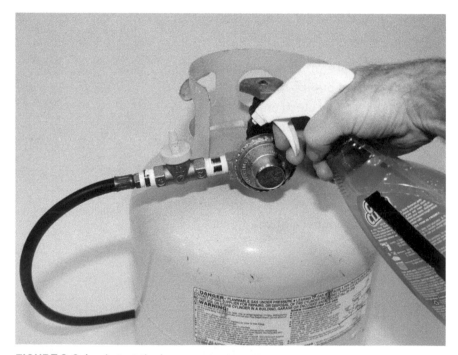

FIGURE 3-9: Leak-test the low-pressure source

5. Close the cylinder valve all the way and open the ball valve to vent the line.

That's all there is to constructing the low-pressure source. We'll use this source with the flambeau, fire pit, and Rubens tube projects.

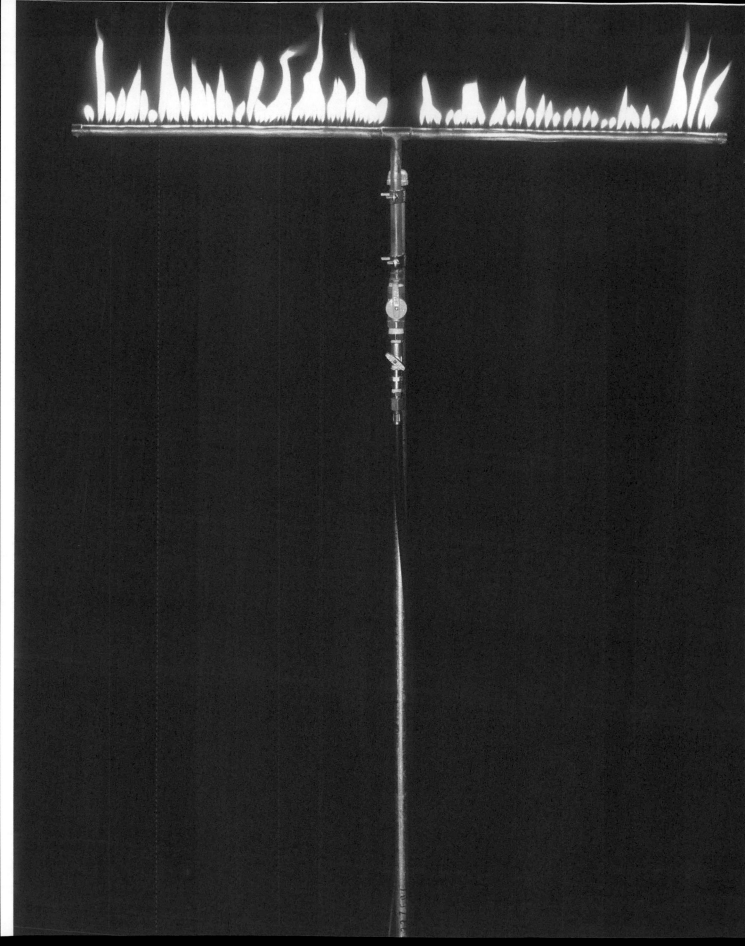

4

The Flambeau

IF YOU'VE EVER BEEN in New Orleans for Mardi Gras and seen one of the night parades, you've probably seen *flambeaux* (the plural of *flambeau*). These are the torches carried in front of the floats to light the way. While some are still burning liquid fuel, many are gas powered. A propane-fueled flambeau would consist of a fuel reservoir, a regulator, and piping to carry the gas to a burner. The traditional flambeaux are portable and human-carried. You may decide to build one like that, but our project is going to be a stationary flambeau suitable for lighting a lawn space.

The basic idea of burning propane for light can take many elaborate forms. In the next chapter we'll discuss bending the burner into various shapes so that it could be a heart, a peace sign, or anything else you can construct. With enough effort, you could spell words or draw pictures! Additionally, many incredible effects have been done using a simple burner and a mask in front so that the flames backlight the cutout pattern. That cutout could be letters, a design, more flames, or a logo of some sort.

Many theme camps in the Burning Man community have a flaming arch at the entry. This is another use for the basic flambeau concept. While this project is fairly simple, there's no limit to how creative you can get with it!

Below is the schematic diagram of a flambeau. (See Figure 4-1.) It's pretty basic.

FIGURE 4-1: The flambeau's schematic diagram

Parts

On the opposite page you can see our parts list for the basic flambeau. The flambeau you want to make may require a different configuration than the one described. Example changes could include reversing the order of the ball and needle valves, adding a propane-rated quick connect at the flambeau, or driving multiple flambeaux from a manifold. These changes would require a different set of fittings and couplers than the ones listed in the table. Using different configurations of parts is okay as long as you understand the substitutions you make.

Safety Note

Always consider the following when substituting parts:

Is the part rated for the pressure required? Remember, to have a safety margin, this has to be higher, ideally many times higher, than the pressure you intend to work at.

Will the part leak? Are the fittings appropriate for propane use (flare-to-flare, taped pipe threads, etc.)? Stay away from the dreaded hose barb!

Will the part degrade in the presence of propane? The main causes for concern here are rubber (latex) or silicone. Both of these are common materials for seals in compressed air parts, even in acetylene welding hoses, and are not safe for use with propane. Use parts with propane-rated Buna-N (nitrile), Viton, or PTFE seals.

Are you matching the materials correctly? See the "Fittings" section in Chapter 2 that describes which pipe materials are allowed to be connected to which fitting types. As an example, black iron pipe connecting directly to copper is not code compliant.

Flambeau parts:

REF	ITEM	QTY
F1	Brass adapter ¼″ MIP × ⅜″ MFL	1
F2	Brass bushing ½″ MIP × ¼″ FIP	1
F3	Brass coupler ¼″ FIP × ¼″ FIP	1
F4	Copper adapter ½″ cup × ½″ MIP	1
F5	Copper cap ½″	1
F6	Copper tee ½″ cup	2
V1	Gas-rated ball valve ½″ FIP × ½″ FIP	1
V2	Needle valve ¼″ MIP × ¼″ FIP	1
P1	Copper tube ½″ × 6′ (to be cut)	1

Stand parts:

REF	ITEM	QTY
S1	Galvanized iron pipe ½″ × 6′	1
S2	Conduit hangers ½″	4
S3	Bolts ¼″ × ½″	2
S4	Nuts ¼″	10
S5	Lock washers ¼″	10
S6	Galvanized floor flange ½″	1
S7	Rebar ½″ × 4′	1
S8	Bolts ¼″ × 1½″	4
S9	Plywood sheet ½″ × 24″ × 24″	1
S10	Beam clamp 1″ × 1¼″	4
S11	Bolts ¼″ × ¾″	4
S12	Galvanized ½″ pipe cap (optional)	1

The block diagram in Figure 4-2 represents how the parts are assembled:

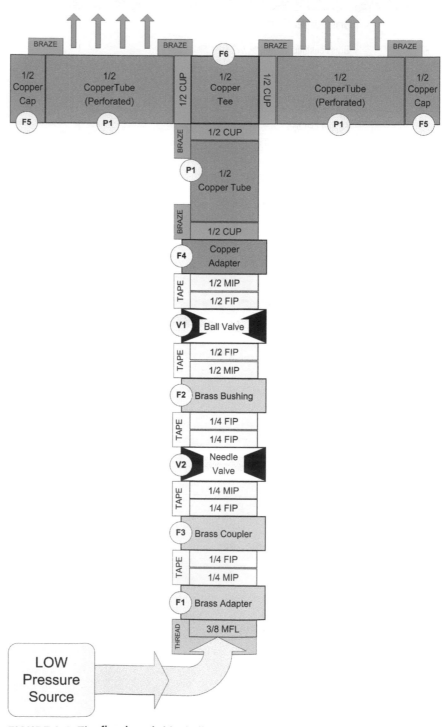

FIGURE 4-2: The flambeau's block diagram

Brazing

Most residential plumbing repairs are performed with soldering, but that's not an acceptable method of joining parts to be used in high-pressure propane systems. NFPA 54, The National Fuel Gas Code, section 5.6.8.1 states that if you're brazing nonferrous metals (e.g.,copper) the filler metal has to have a melting point in excess of 1000°F (538°C) and have no more than 0.05% phosphorus. This means that many of the common brazing rods that contain phosphorus in amounts of 5% or higher wouldn't be approved.

Copper tubing, according to the NFPA 54 National Fuel Gas Code, must be brazed, not soldered. This is to ensure that the joint can withstand the required pressures, and can withstand extreme temperatures in case of a fire, without venting gas.

As mentioned earlier, in Chapter 2, "Equipment and Parts," many authorities having jurisdiction (AHJs) find it difficult or impossible to permit brazed joints without expensive tests and documentation. Burning Man is an important example of this. Out in the Black Rock Desert, trying to keep tens of thousands of people safe, their approval process for flame effects has to uphold extremely high standards. Before planning to bring a flame effect you've constructed to an event, research the standards published by the AHJ. If you have questions, contact them and ask for clarification.

While many AHJs, such as Burning Man, have a single standard regarding brazed fittings, many AHJs will permit brazing for nonpressurized fittings. Based on this, and the value of brazing as a skill, I'm going to proceed with it for the projects in this book. The only other option is to construct burners out of black iron or steel pipe (which are a nightmare to bend) or to use flared fittings at all joints and caps. If your AHJ does not permit nonpressurized brazed joints, you will likely have to make the choice between iron or steel burners or flaring. In any case, soldering is an unsafe option for burners or flaring due to the low capacity to withstand heat.

Brazing and soldering both use a filler metal to join parts. Brazing is different than soldering by virtue of the temperature at which it occurs. Below 840°F (450°C) you're soldering; above that you're brazing. Because of the difference in melting points, the filler metals used for brazing and soldering are also different.

Be smart: Brazing is a high-heat operation so you want to perform it in a safe location with good ventilation. Inspect the area you're going to use and imagine how you're going to move around while holding the lit torch. Visualize where the hot parts are going to be.

If they have a chance of falling, are they going to go somewhere undesirable—carpet, bare feet, into a trash bin full of paper? If molten brazing alloy drips off the part while you're brazing, what will it fall on? As you move around the part, are you going to knock into anything? Is there any chance you're going to try to grab the sizzling hot part unexpectedly? (Actually, it's not the *part* that sizzles if you grab it, so really, *really* don't grab a part with bare hands or inadequate gloves.)

One or more firebricks can be extraordinarily useful for operations like brazing. They can serve to protect surfaces from the torch or dripping brazing alloy. They can serve to help position the parts for brazing. Work clean; firebricks tend to be dusty and you don't want to get that dust on your clean or fluxed joint.

If you read up a lot on brazing you'll probably come across various descriptions of how brazed joints require less penetration than soldered joints. The copper fittings purchased from local hardware stores are designed for soldering, and technically position the fittings too deeply into each other for the perfect brazed joint. However, you may also notice that the perfect brazed joint can be rated as high as 135,000 psi. This leaves us a considerable safety margin to use common fittings and position the parts normally within them. If you're really interested in brazing, you can purchase a special tool that will create an optimal brazing detent in your fittings. A *detent* is a small dent you make that will help you determine something by feel. Sometimes knobs that have a center position will use one to let you know when you've reached it. The detent provided will position parts at the perfect penetration and not allow them to go any deeper. But most sites recommend, and we can get by with, using standard fittings for our projects.

There are lots of excellent YouTube videos that show how to braze copper pipe. Taking the time to watch a few of them will be a big help in visualizing the steps to undertake.

Tools

There are just a few additional tools required for this project.

FIGURE 4-3: Triangle file

TRIANGLE OR SQUARE FILE   (See Figure 4-3.) This file has a sharp edge that we'll use to notch the copper tubing and make it easier to pierce. Files often are easiest to buy as a set. Most sets will include a triangle or square file. Buy a full-sized file, 8″ (200 mm) or more.

WIRE NAIL   (See Figure 4-4.) Any kind of small wire nail; brad; or tiny, sharply pointed, nail-like item will do. The purpose is to make a very small hole. I like to use #18 wire nails or smaller.

SANDPAPER   (See Figure 4-5.) A pack of mixed-grade grits will cover all our needs.

A DRILL KIT   (See Figure 4-6.) Consisting of:

DRILL   A dependable corded or cordless drill is a workhorse tool. There is no tool I've used harder or replaced and upgraded more often than my cordless 18V drill. But I still pull out my 35-year-old corded drill on a regular basis. If you're not ready to spend the money for a cordless drill and good batteries, there are great deals to be had on plug-in models and an extension cord.

DRILL BITS   You'll need a set of drill bits. A small set will do. Because of the next item I'm going to suggest, you don't need every imaginable size (though it's odd how often I use some of the rarer tiny numbered bits). The titanium-coated ones are great, but not absolutely essential.

FIGURE 4-4: Wire nail

FIGURE 4-5: Sandpaper, aka *emery cloth*

FIGURE 4-6: Drill, drill bits, and step drills

STEP DRILLS Step drills allow you to make a large number of different-sized holes with a single bit. It's also much easier to drill a large diameter hole in steps rather than all at once. The drawback to step drills is that they can only make a relatively shallow hole (¼″ or about 6 mm). This isn't a problem as often as you might think, since a lot of holes are for mounting things in material that isn't very thick.

Construction

1. Cut the Copper Tubing

While the flambeau can work successfully in an endless series of configurations, we're going to start with a simple T. Arguably, it would be slightly easier to skip the crossbar and just make a straight tube with a cap at one end and the inlet on the other. But the T configuration has a few advantages.

The first advantage is that, by placing the inlet at the center of the crossbar, we distribute the gas more evenly. A simple tube might not show the problem clearly, but in larger systems the pressure can vary with the distance from the inlet. This can cause the flames to be higher at one end of the burner than the other. While it's possible to mitigate this by slightly increasing the size of the jets (holes) as you progress away from the inlet, we'll address the problem of pressure distribution by placing the inlet in the middle.

The second advantage is that the riser (the shaft of the T) provides us with a mount

point that allows us to handle the burner without directly grabbing the open flame. Additionally, it provides some separation between the inlet valve and the hot burner. However, keep in mind the riser can still get hot! It's just not "open flame" hot.

Note that I'm going to use slightly different dimensions for the metric version, so the conversion won't be exact.

Our T will have two 18″ (400 mm) burner arms on each side and an 8″ (200 mm) riser. Copper is expensive, but since I always have a use for it, I purchased a 10′ (3 m) length of ½″ (12 mm) tubing. Unless you happened to have exactly a 42″ (1 m) section of tube, or chose different dimensions for your crossbar and riser, we'll need to make three cuts. We'll start with the 8″ (200 mm) riser by measuring and marking our cut line. (See Figure 4-7.)

We'll use a tubing cutter to make our cuts; if you prefer using a hacksaw, that will also work. However, when used correctly, tubing cutters make clean precision cuts. It's surprisingly easy to create jagged edges with a saw. I've come to prefer tubing cutter cuts, so we'll use them in this project.

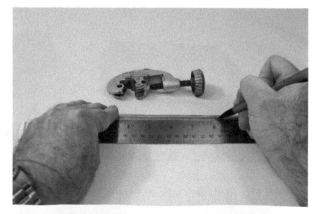

FIGURE 4-7: Mark the cut line on the copper tube.

Open the tubing cutter with the bottom knob so that it fits over the tube. Align the circular blade so that it is directly on the cut line. Hold it there carefully while tightening the adjustment knob. Be careful; the tube must be held between the two rollers at the bottom and by the blade at the top. Do not overtighten. Cutting tubing is one of those things that works with a series of small adjustments rather than brute strength. Tighten just enough so that, as you turn the cutter around the tubing, you see a light score line. (See Figure 4-8.)

FIGURE 4-8: Cut the copper tube with the tube cutter.

Here's an important tip: every time you tighten the knob (less than a quarter turn), do it at a different point around the circumference of the tube. It's natural to want to tighten at the same point every time, but this will end up causing a slight deformation in the edge of the tube. If you're not careful, by the time you're done it can be difficult to fit the tube into a fitting and it may be prone to leakage.

Occasionally, you will see the cutter execute a spiral score on the tube instead of lining up in a tight circle. This is typically due to a loose cutting wheel. Tighten the screw holding the cutting wheel in place and carefully keep it aligned to make a clean circle. Avoiding a deep spiral cut is another reason to make your first pass a light score.

Rotate the cutter around the tube two to three times until it moves easily again.

Each time you tighten (again, less than a quarter turn), rotate the cutter around the tube two to three times until it moves easily around the tube. You'll build up a rhythm pretty quickly and cut through the tube in no time.

The edge of the tube will be bent inward, forming a lip that restricts the diameter of the tube. (See Figure 4-9.) The size of the lip depends on how sharp the cutting wheel was and how hard you turned the knob. Nothing is free; if you tried to save time earlier by taking a deeper bite on each tightening cycle, you'll pay for it when you ream the tube to restore the passage. It may be tempting to skip reaming the tube, but if you don't, it will leave a poor edge on the tube and reduce the effectiveness of our join.

FIGURE 4-9: The cut tube has a flanged lip.

To use the reamer built into the tubing cutter, push the blade gently into the throat of the tube and turn. This only works easily with small, thin cuts. If the blade bites deeply into the inner lip, turning it will be very difficult and you will end up making an uneven, jagged set of cuts. I will not lie to you; I hate the reamer that comes on the tubing cutter. It's better than nothing, but if you use a dedicated deburring reamer, you will never willingly use the built-in one again. The conical deburring reamer has three blades and fits smoothly into the tube, turning in an ergonomic manner. I know it's one more thing to buy, but if you do much copper cutting you won't regret having one.

Proceed to measure, mark, cut, and ream the remaining two 18" (400 mm) pieces in the same fashion.

2. Mark the Jets

Lay out and dry-fit the parts so you can spot any mistakes. (See Figure 4-10.) Hopefully everything looks good, but if not, go back and redo whichever piece is a problem.

Fit each tube into the ½" (12 mm) copper tee and mark the tube along the edge of the

tee with a pencil line. Do the same with the cap on the other end. The space between these two marks is where the jets are going to be made. (See Figure 4-11.)

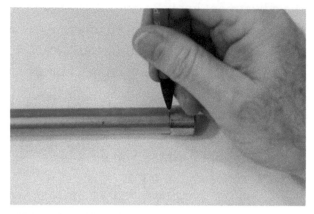

FIGURE 4-11: Mark the tubes where the cap and tee line up.

You can place as few or as many jets as you like. I'm going to place them ½" (12 mm) apart, allowing for 33 jets on each side.

Take a ruler and tape it to each end of the tube so that its center is evenly spaced between our two previous marks. Mark a pencil line along the length of the tube where the ruler's edge touches. This is the lateral line where the jets will be placed. (See Figure 4-12.)

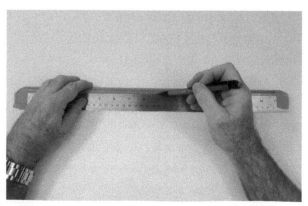

FIGURE 4-12: Mark a line along the tube to line up the jets.

FIGURE 4-10: Lay out the parts to visually check the fit.

Make pencil marks on the tube at the appropriate distances on each side of center. If your ruler was too short, make a set of marks in one direction, then reposition and retape the ruler using the previous marks as registration. These radial marks, perpendicular to our initial mark, give us the position of each individual jet. (See Figure 4-13.)

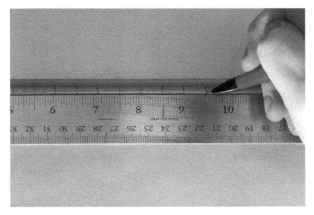

FIGURE 4-13: Make marks for each jet.

3. Punch the Jets

If you have a drill press and a pipe mount, you can drill each of the jets using a tiny drill bit (like a #56 bit). However, even with a good drill press, I prefer to do the next step using a different method.

Gently clamp the tube with a vise (don't distort the circular shape) or hold it firmly in your hand. Pick a jet to start with and, using the edge of the triangular or square file, lightly score a line tangentially across the tube at the point where the lateral line and radial mark cross. Keep the file tangential to the tube at the lateral line. Gently continue to file the groove until you go most of the

way through without creating a hole. Don't worry if you are slightly off or accidentally mar the surface when the file slips. These marks will clean up or cover up over time. (See Figure 4-14.)

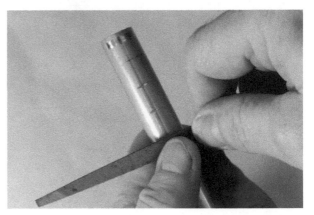

FIGURE 4-14: File a groove where each jet will go.

If you do accidentally go all the way through, we can fix it. Try not to go too deep. If the hole is tiny enough, it may serve as an effective jet all by itself. Most likely it's smaller than the jet hole we're going to punch. In the unlikely case that it's too big, you can braze over all or part of it later.

This is a good time to reestablish the lateral pencil line. This will usually leave a mark in the groove at the point where you will want to punch the jet. Take a wire nail and a hammer and lightly rest the point of the nail in the groove along the lateral line and gently tap it with the hammer to barely pierce the copper. We can come back later and enlarge it if needed. Only the very tip of the nail should go in if you're using anything larger than a #18 wire nail. If you accidentally hammer it all the way in, pull it out; we can braze over it later if it's too large. If you used a #18 or smaller nail, hammering it all the way in is fine if you pull it out carefully. (See Figure 4-15.)

FIGURE 4-15: Punch a hole in each groove to serve as a jet.

Even though brazing could cover up the holes we don't want, the brazing filler is much harder than copper and will be considerably more difficult to file or pierce. Piercing the copper directly is much easier. If you don't feel confident, try a few practice grooves and piercings on a scrap piece of tube to get a feel for it.

Proceed to groove and punch the rest of the jets at each of the markings. If you're feeling the urge, go ahead and add one to two jets to the top of the copper tee (but avoid the overlap between the edge of the tee and the inserted end of tube).

4. Clean the Parts for Brazing

The goal in this step is to remove any grease, dirt, oxides, or other material that might hinder the flux or brazing material from flowing evenly. Use fine-grit sandpaper, a wire brush, or emery cloth around the ends of the copper tubes and inside the cups of the fittings. The copper should be bright and shiny. (See Figure 4-16.)

If the parts are oily or have crud on them, a good degreaser (carb cleaner, brake cleaner, engine degreaser, etc.) should clean off any of the gunk. Don't forget to clean inside the cups of the fittings as well as the ends of the tubes.

If you're going to use chemical degreasers, do so outdoors or someplace with excellent ventilation. Wear eye protection (always!), especially if you're using a spray can. Personally, I always put on a pair of nitrile gloves when working with stuff like this. We'll be using disposable nitrile gloves in the Rubens tube project later on, so you may want to buy yourself a box of them. You'll find they come in handy in a lot of situations.

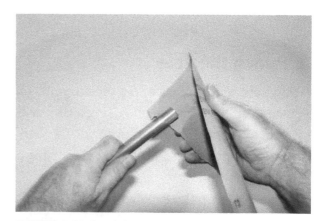

FIGURE 4-16: Sand the copper until it shines.

5. Braze the End Caps onto the Burner Tubes

If this is your first time brazing, I highly recommend buying a couple of extra copper caps and a short length of tube you can practice with. Cut the tube into a few short sections and go through the following steps to braze the caps on. A couple dollars' worth of sacrificial material will be well worth it if you'd like to feel confident before attempting to braze something you really care about.

Dry-fit the cap onto the tube to determine how much of the tube the cap covers. Remove the cap and use the flux brush to apply a thin layer of flux to the end of the tube just beyond where it penetrated the cap. (We marked this in step 2.) Apply a thin layer of flux inside the cap, as well. (See Figure 4-17.)

Position the cap fully onto the end of the tube and give it a slight twist to even out the flux. Make sure the tube is lightly clamped in a vise or otherwise held securely in place. You can rest the tube on a fire brick, with the cap and an inch or so hanging off the end, if you place something to keep the tube from rolling. The tube will get very hot where you're brazing, but won't be as hot 12–14″ (30-35 mm) farther down. Pay attention to what will be "in the line of fire" of your torch while you're heating the tube. I consider

most of the scorch marks on my workbench badges of honor, but you may feel different about the surface you're working on.

Put your gloves and safety glasses on, light the torch, and begin heating the tube just below the cap. (See Figure 4-18.) Heat the tube primarily from below and let the heat rise through the joint rather than from the top down. You also don't want the torch to remain still and overheat any one spot. Move the torch over and around the tube and cap in a short swirling motion to evenly heat the joint. The copper will change color and the flux will serve as a useful indicator of the temperature. Keep heating until the flux ceases bubbling and becomes calm and clear like water.

When the joint is ready, move the flame to the overlap of the fitting and press the tip of the brazing rod to the edge of the joint (the lip of the cap). (See Figure 4-19 on the next page.) While you normally avoid playing the flame on the brazing rod, occasionally it can be helpful to briefly aim the torch's flame at the tip of the rod to start it melting. However, a properly heated joint should generally

FIGURE 4-17: Applying flux to the parts

FIGURE 4-18: Heat the parts.

start the rod melting by itself. As it melts, move the rod around the rim so that it is drawn under the cap all the way around and into the joint.

It's important to remove the flux residue after brazing (it becomes corrosive over time). (See Figure 4-20.) A brisk scrubbing with a wire brush and a wipe with a dry cloth will usually do the trick, but a warm wet cloth will loosen and wipe off whatever flux is left over. Some folks like to quench the joint in hot water to crack the remaining flux, but this isn't always feasible.

Pro tip: Be careful when using a plastic-handled wire brush to scrub the hot tube or fitting. If the handle slides across or

FIGURE 4-19: Applying filler

FIGURE 4-20: Cleaning the joint

bumps the tube, you'll get a streak of melted plastic. Use wood-handled wire brushes if you can find them.

Here are a few ways to troubleshoot common brazing problems:

- If the filler doesn't flow, or balls up, the parts aren't fully heated or there is still oxidation on them.

- If you see the parts oxidizing (turning black) while you're heating them, there isn't enough flux.

- If the filler doesn't enter the joint, or just flows onto one of the parts and not the other, then the parts aren't evenly heated.

If you're like me and braze infrequently, you may not be happy with the way it looks; for example, globs of filler, chunky parts, or uneven brazing marks may make you grimace. I offer the following advice:

- The flambeau is likely to be shooting flames and be positioned about 6′ in the air. People aren't going to get as close of a look at it as you might think.

- A file or small grinding wheel can rectify a multitude of sins. Grind or file offending filler away; just don't cut into the copper.

- You can reheat and clean the joint up. Use the tang of a file or some other heat-resistant metal device to smooth things out when they're liquid. You can also reposition off-kilter components when the joint is reheated.

- You can always heat it back up, pull it apart, file (more likely aggressively sand) it down, and try again. This really isn't a great strategy, but if all else is unacceptable, doing this or just starting over fresh is always an option.

The rest of the brazing activities we'll undertake in this book go through these same steps. Go ahead and braze the cap on the other tube so that both burner tubes are ready to braze into the tee.

6. Braze the Burner Tubes into the Tee

Our next step is to braze the two burner tubes into the arms of the copper tee.

Clean the tube ends and insides of the tee arms, as described previously. Mounting the base of the tee in a vise with the joints accessible for brazing is really the very best solution for holding the parts while brazing. Flux the joining surfaces and mount the tubes into the tee, twisting them a little to even out the flux in the joint. Be careful that the jets are pointing up when you finish. The tubes should fit firmly in the tee, but if they sag, you'll need to support the ends somehow. If nothing else comes to mind, you can bend a couple coat hangers and use them as supporting arms attached to the vise. Anything that can support them will work—chairs, boards, tables, or whatever. The ends will get hot, but not scorching hot. An assistant could hold the tube in position with pliers or welding gloves, but probably only needs to do so at the very last to orient the tube as you pull the torch away and the filler hardens. (See Figure 4-21.)

Braze both tubes in place. We're going to braze another piece right away, so unless

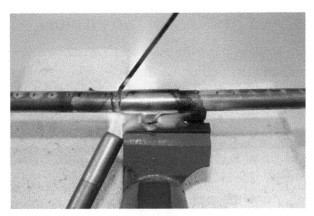

FIGURE 4-21: **Brazing the tubes**

you're taking a long break at this point you can defer cleaning up until the next step's done.

7. Braze the Riser Arm and Fixture

Clamp the riser tube in the vise and attach the tee. Horizontal brazes are easier than vertical ones. Clean and flux the riser tube and remaining fitting on the tee. Position the riser tube in the tee and braze it in place. (See Figure 4-22.)

The last component to braze is the pipe thread to cup adapter. Clean the end of the riser tube and interior of the cup fitting, flux, and braze. Clean up the brazes on the tee and adapter.

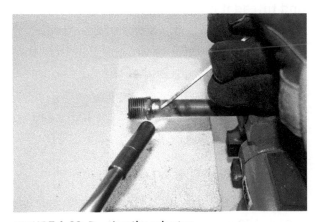

FIGURE 4-22: **Brazing the adapter**

8. Add the Rest of the Fittings

The rest of the construction consists of connecting the additional fittings. I tend to refer to a bunch of connected fittings as a *stack*. In each case, except for the flare fitting at the very end, we'll use yellow Teflon tape to wrap the threads and make them gas tight.

In order, starting from the copper adapter, the stack consists of:

- ½" ball valve (V1)
- ½" MIP × ¼" FIP brass bushing (F2)
- Needle valve (V2)
- ¼" brass coupler (F3)
- ¼" MIP × ⅜" MFL brass flare adapter (F1)

Begin with the copper adapter; tape its threads and attach the ball valve. You will need two crescent wrenches to get this, and subsequent parts, tightened correctly.

Next up, tape the end of the ½" MIP × ¼" FIP brass bushing and thread it into the other end of the ball valve. Follow by taping the remaining end of the reducing nipple and thread it into female end of the needle valve. Tape the male end of the needle valve and thread it into the coupler. Tape the threaded (not flared) end of the flare adapter and thread it into the coupler.

That's it! Construction of the flambeau is complete. (See Figure 4-23.)

Constructing a Stand

There's still the question about how to safely mount the flambeau. This largely depends

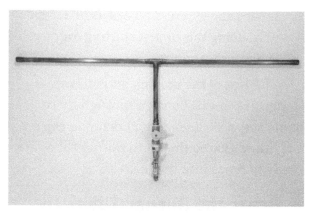

FIGURE 4-23: The completed burner

on your intended use, but there are some important concerns in every case. First and foremost is the need to avoid a large flaming burner falling over and inciting mayhem and injury. The environment where the flambeau will operate is a critical aspect. Is high traffic anticipated? Is there a chance that the propane cylinder or hoses will be handled or knocked around by people? Is the ground covered in flammable dry leaves? Are the controls accessible when the flambeau is in position?

I'm going to describe a simple stand that will work for many, but not all, situations. Please take it as a starting point for your own ideas.

The primary clamp point for the flambeau is the riser tube. Hose clamps, pipe hangers, and pipe straps are all useful components for attaching something like the flambeau. Prowl the plumbing and electrical aisles of the hardware store. EMT conduit hangers are really useful parts for a lot of mounting situations; we'll be using them extensively.

To construct the basic stand, we'll mount the flambeau to the end of a length of galvanized iron pipe. Black iron pipe would work fine, but since this is intended for outdoor use, galvanized pipe avoids rust. If you decide to use black iron, consider painting it to keep it from rusting. In our example we'll use a 6' (2 m) pipe so that the controls are still within reach. The pipe is attached to a galvanized floor flange on a 2'×2' (approximately ½ m²) sheet of plywood. We'll put a pipe cap (S12) on the top of the pipe to keep water out; this is optional but keeps things looking nice. The plywood provides space to mount the cylinder. The stand is stabilized with a 4' (1 m) length of rebar driven halfway into the ground and up through a hole in the plywood, through the flange, and into the pipe. The rebar offers a lot of stability, but won't stop a concerted effort to knock the flambeau down (I have a lot of rowdy friends, so despite my emphasis on safety, this kind of problem keeps coming up in my world). If you need extra stability, drive tent stakes around the edges of the plywood, and if the stakes don't hold the edge directly, tie ropes from stake to stake across the plywood. In most cases without traffic, heavy wind, or mayhem, the base and rebar will be just fine.

Connect two of the ½" conduit hangers (S2) base to base with a ¼" × ½" bolt, nut, and lock washer (S3, S4, S5) so that the openings for the tube are opposing each other and on the same level. (See Figure 4-24.)

We'll use this bracket in a number of the projects; it's an easy way to mount pipes to each other.

Remove the tightening screw on one of the hangers and pull the mouth open

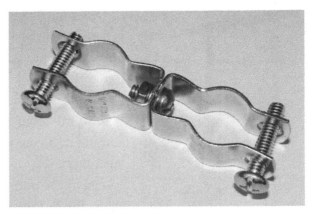

FIGURE 4-24: **Conjoined hangers**

enough to be able to push over the copper tube. Push it onto the tube, replace the tightening screw and tighten it all the way until it makes a firm mount. (See Figure 4-25.)

Do the same thing for the side that mounts to the end of the galvanized iron pipe (S1) and attach the flambeau to the pipe. (See Figure 4-26 on the next page.)

Position the floor flange (S6) and the propane cylinder on the plywood (S9) so that they are evenly placed. The diagram in Figure 4-27 provides some suggestions for the

FIGURE 4-25: **Attaching hanger to the flambeau**

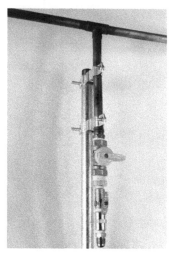

FIGURE 4-26: Attaching flambeau to the stand pipe

dimensions. They're not critical in any way; as long as everything is firmly attached and the flambeau doesn't fall over, it's all good. Mark the board where the bolt holes and center hole of the floor flange are.

Drill the ¼" bolt holes for the flange and the ½" center hole. Test fit the rebar in the center hole; it should fit through without forcing it. If you use a step drill bit you can easily enlarge it if necessary.

Push the bolts (S8) through the bolt holes from the bottom. Place the floor flange onto the bolts, put lock washers and nuts (S4, S5) onto the bolts, and tighten.

We're going to repurpose Unistrut (aka *Superstrut*) beam clamps to mount the propane cylinder to the plywood. The base rim of the cylinder will sit in 4 of these and the clamping bolts will go through the open holes in the rim. The rim on

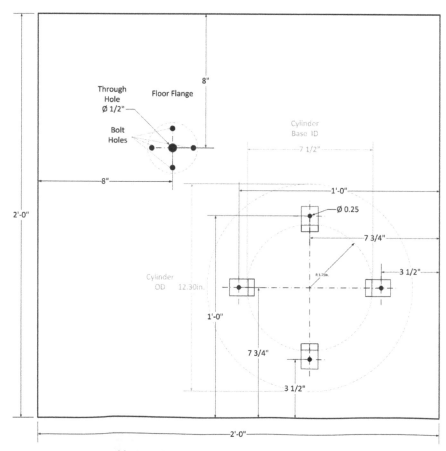

FIGURE 4-27: Stand layout dimensions

the base of the propane cylinders takes tremendous abuse. It is not uncommon for them to be bent out of true. Use pliers or vise grips to bend the rim so that it will fit in the mount if necessary.

Drill four ¼″ holes in the base at the places shown in the diagram. The beam clamps have a mounting hole threaded for ¼″ bolts. Push a ¼″ × ¾″ bolt (S11) through from the bottom and thread the clamp onto it until it is tight and the locking screw is facing outward.

The completed base should look something like Figure 4-28.

The cylinder sits in the clamps so that the back edge is on the inside of the rim and the holes in the rim line up with the locking screws. (See Figure 4-29.)

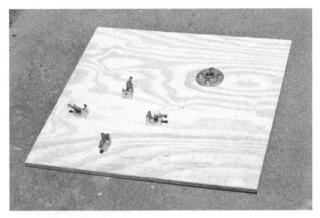

FIGURE 4-28: **The completed base**

FIGURE 4-29: **Cylinder positioned in clamps**

Testing

To assemble the stand for use, hammer the rebar halfway into the ground and place the plywood over it so that it protrudes through the floor flange. Position the cylinder so that the base rim fits into the beam clamp mounts and the oval holes in the base rim are centered where the clamp's bolt will go through them. Tighten the beam clamp bolts completely.

Slide the pipe, with the flambeau attached, over the rebar and thread it into the floor flange. Attach the gas-rated hose to the flare fitting on the flambeau and verify that the ball valve and needle valve are closed. (See Figure 4-30.) Verify that the ball valve at the regulator is closed and attach the regulator to the propane cylinder. Put your safety glasses on.

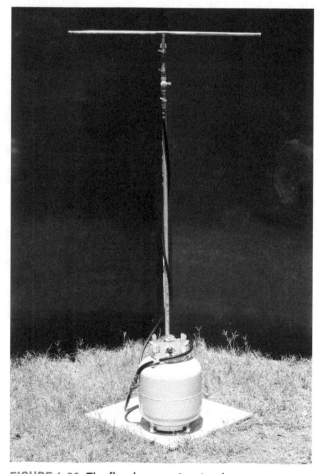

FIGURE 4-30: The flambeau on its stand

Open the cylinder valve all the way and then back it off half to three-quarters of a turn. Test for leaks by spraying the connection at the cylinder and at the ball valve (see Chapter 3, "The Low-Pressure Source," for instructions on making a leak testing kit). You don't have to soak it; just make sure you spray on a light, even mist. If bubbles emerge, close the cylinder valve completely, tighten the leaking joint, and repeat until no bubbles emerge.

Open the ball valve nearest the regulator and leak-test the section between that ball valve and the needle valve on the flambeau. Tighten any joints that bubble.

Open the needle valve and test the section up to the final ball valve. Tighten if bubbles emerge.

The sections that will be under pressure have all now been tested! The last remaining step is to light it up. Once we open the final ball valve, propane will begin flowing out of the jets on top of the flambeau. We want to avoid discharging unburnt gas as much as possible (remember, it sinks, so be careful of ignition sources at ground level). There are a number of ways to light the flambeau, the most convenient of which is probably a fireplace lighter. (See Figure 4-31.)

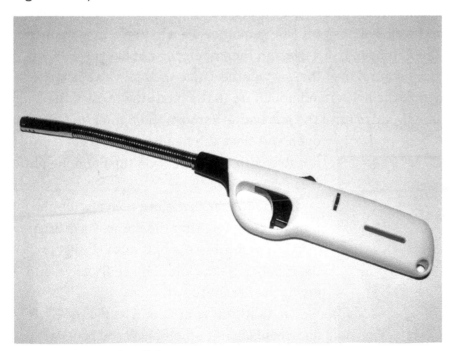

FIGURE 4-31: Fireplace lighter

Fire Extinguishers

Before we involve fire in our testing, make sure you're prepared for surprises or accidents. A garden hose is a great tool to have at hand. A fire extinguisher is also an important piece of safety equipment. However, if you've chosen a commonly available ABC dry chemical (chem) extinguisher, and you've never had the opportunity to discharge one, be aware that they make a *big* mess. The yellow powdered chemicals they spray (in shocking quantities) are not something you want to ingest or leave in contact with your skin. They can also be corrosive to metal over time. Using a dry chem extinguisher is far better than having an out-of-control fire, but the stakes have to be pretty high before I will discharge one.

Alternatives exist. Both pressurized water and carbon dioxide (CO_2) extinguishers can be safely used around people and animals (but be careful because CO_2 can be very cold). Unfortunately, both are relatively expensive and difficult to obtain compared to dry chem. This is why I want to emphasize the value of garden hoses, wet blankets, and sand as firefighting tools for the propane experimenter. They may be old school, but in many situations, they're a great choice. Use good sense and, if possible, have multiple levels of response, saving the dry chem for when it's really needed.

To avoid venting gas at full pressure, turn the needle valve to its closed position. The sequence we will perform is to open the flambeau ball valve and then open the needle valve so we have control over the gas volume. Once the ball valve is open, ignite the lighter and hold it up to the jets. Slowly open the needle valve until the jets ignite. Remove the lighter and slowly open and close the needle valve to get comfortable with the impact on the flame height. (See Figure 4-32.)

To shut down, first close the valve on the propane cylinder. After the flames have extinguished, then close the ball valve closest to the cylinder. Next, close the needle valve, and finally the flambeau ball valve.

That's it! You've built and tested the flambeau! Light it up, sit back, and bask in its gentle glow.

FIGURE 4-32: Lighting the jets

Safety Note

The flambeau is a burner. It generates a great deal of heat. Pay close attention to what is above the flambeau and make sure nothing could scorch or catch fire. Do not operate this burner indoors. The copper tubing will also get extremely hot. Once lit, or after it's been lit, do not handle it without heavy leather gloves or pliers. The longer it burns, the hotter it will get. Pay attention at various intervals to how hot the ball valve is getting. If it becomes too hot to touch, shut the system down per the previous instructions.

Ideas for Enhancing

Weatherproofing

The easiest, and possibly most useful, upgrade is to seal the plywood you used for a base. If you're intending for the flambeau to live outside in the weather, making sure the wood doesn't warp or rot is a good idea. I'm a fan of Thompson's® WaterSeal® or similar products. A good coat of paint is also an option. For best results, take all the fittings off first before sealing or painting.

Earlier, I recommended using galvanized pipe and floor flange for the same reason. Exposed sections of black iron pipe will rust in a surprisingly short amount of time. If you do use black iron pipe, I absolutely recommend painting it. I'll talk more about the benefits of this in Chapter 9, "The Boosh," in the section on building the boosh, but for now let me say that painted black iron looks dramatically better than rusted black iron.

Even Out the Jets

It's likely that you will notice that the flames coming from the jets are of different sizes. This can be remedied by using one of the same wire nails you used in the steps earlier. While pushing the nail all the way in on every hole is the most consistent method of making the jets the same size, you'll still possibly notice that the jets closest to the feed are slightly larger due to a pressure gradient. Perfection is difficult to achieve without careful fabrication methods that our handmade efforts have trouble duplicating. My advice on aesthetics, for what it's worth, is to get it close to what you like and enjoy some wabi-sabi (the Japanese concept for the appreciation of the beauty of imperfection). In the end, air currents around the burner will impact the size of the individual flames as much as anything you can do to standardize them. Fire and flame are visual expressions of chaotic influences and, I believe, have a lot of beauty in their randomness.

Provide Air Mixture

You may also notice that the flames rising from the jets are accompanied by black smoke. Chapter 1, "Understanding Propane," provides the explanation for this and a description of what is being produced by the combustion. The short answer is that insufficient oxygen is mixing with the propane to achieve complete (stoichiometric) combustion. If you took chemistry in high school, you may recall using a Bunsen burner. The difference between our burner and a Bunsen burner is the introduction of air into the mix prior to ignition. Burners on propane grills do the same thing.

We could introduce holes into the base of the riser tube (the base of the copper tee) that would suck up air and mix it with the propane. Various methods to cover selected holes to moderate the air input are relatively easy. A sliding sleeve around the tube with a hole that overlaps the one in the tube is a common method. Aluminum ducting tape or even hose clamps could achieve a similar effect.

The downside, for some purposes, is that the flame will get considerably hotter when mixed with air. When using the flambeau for ornamental lighting, additional heat can be undesirable. If you do choose to introduce air, I recommend cutting the riser tube and brazing in an additional section of tubing to provide a longer thermal barrier to the controls. This has the added advantage of making it possible to add a sleeve for air control.

Front with a Metal Mask

The prettiest effects from the flambeau, in my opinion, are when the flames backlight metal cutouts. Metalworking is a different book, but a pair of aircraft tin snips and some sheet metal (find the cheap sheets in the ductwork section of the building supply store) or corrugated metal can make some great-looking effects. Be careful and don't cut yourself!

The Portable Fire Pit

IN THE PREVIOUS CHAPTER, we constructed a basic low-pressure burner. With small modifications that burner can be put to many uses. One of these is the time-honored tradition of sitting around a fire, contemplating the flames. Propane has many advantages for this kind of activity: ease of setup, lack of cleanup, and the ability to control output. By gently dispersing the flames from our burner, we can use our previous skills (and a couple new ones) to build an excellent fire pit.

Many propane or natural gas fire pits are permanent installations. This project is a portable fire pit. It's great for locations that don't allow ground fires. It's designed to be lightweight, while remaining extremely stable. Other than the lava rocks we'll use, the whole thing weighs just a couple pounds. It's easy to move or break down and take camping. The concepts in the project could be made into a permanent installation, so don't feel confined by the portable aspects.

The plumbing components for the project are identical to the flambeau. The only difference is the shape of the burner. You can reuse everything from the flambeau up to the copper; just unscrew the brass stack at the point where it connects to the ½" copper adapter. We'll also be reusing the low-pressure source.

The takeaway from this chapter, which will help us in future projects, is learning how to bend copper. In the boosh project, our pilot will utilize a bent copper tubing ring, much like what we'll build in this project.

Many vendors sell special burners for fire pits. These are generally one or more rings with a central feed. For fire pits that aren't round, rectangular, H- and even T-shaped burners are available alternatives. The primary purpose of all these designs is to get an even distribution of gas throughout the pit. DIY and commercial fire pits vary in how much pressure the design is intended for. Introduction of air (oxygen) to achieve a more complete and less sooty burn is also a frequent part of many designs. We'll be using our low-pressure source to provide a small, gentle flame, taking advantage of the irregular surface of the lava rocks to help mix air with the propane.

Online notes from vendors suggest that people crave BIG flames in fire pits. This is generally achieved by using a high-pressure source. This project is *not* designed to support that. Build a heavier installation with fire-resistant materials like concrete or bricks if you want roaring big flames.

For convenience's sake, we'll be feeding our fire pit from the side instead of underneath. You can look at projects online for other design approaches. Our design constraints are to build a safe, working fire pit at the lowest reasonable cost. That doesn't mean you can't improve or change it in any number of ways.

The schematic for the fire pit is identical to the flambeau (see Figure 5-1):

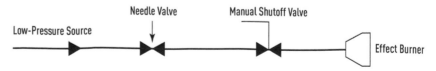

FIGURE 5-1: The fire pit schematic

Parts

The following are the parts needed for the portable fire pit.

From the flambeau:

REF	ITEM	QTY
F1	Brass adapter ¼" MIP × ⅜" MFL	1
F2	Brass bushing ½" MIP × ¼" MIP	1
F3	Brass coupler ¼" FIP × ¼" FIP	1
V1	Gas-rated ball valve ½" FIP × ½" FIP	1
V2	Needle valve ¼" MIP × ¼" FIP	1

Fire pit parts:

REF	ITEM	QTY
P	Straight copper type L tube ½" × 10' (sold in 10' sticks) Cut into P1–P4, as follows.	
P1	Copper tube ½" × 2"	1
P2	Copper tube ½" × 12"	1
P3	Copper tube ½" × 7"	2
P4	Copper tube ½" × 11"	4
F4	Copper adapter ½" cup × ½" MIP	1
F5	Copper cap ½"	4
F6	Copper tee ½"	3
F7	Copper 90° elbow ½"	1
A1	Aluminum hot water heater pan 22"	1
A2	BBQ lava rocks 7 lb bag	3
A3	Mesh strainer for sink drain 2½"	1

Stand parts:

REF	ITEM	QTY
S1	EMT conduit ¾″ × 10′	1
S2	Conduit hangers ½″ (cut into thirds)	3
S3	Bolts ¼″ × ½″	6
S4	Nuts ¼″	6
S5	Lock Washers ¼″	6
C	Double loop chain #3 × 20′ (or a 15′ package and 6′, see text), cut in the following way:	
C1	Double loop chain #3: 29 links	3
C2	Double loop chain #3: 25 links	3
C3	Double loop chain #3: 12 links	3
S6	Quick links ⅛″	6
S7	Turnbuckle hook and eye 4¾″ × 5/32″	3

The block diagram for the fire pit is different from the flambeau in the burner segments (see Figure 5-2).

Safety Note

This project is *only* suitable for use outdoors. Under no circumstances should this be used indoors or in an enclosed space. The propane being burned is only externally mixing with air. This means that the burn is "rich" and may include some small amounts of carbon monoxide (less than a car exhaust). This is fine outdoors, but would not be okay without good ventilation. Additionally, never underestimate the risks with open flame. The fire pit should only be used by, or under the supervision of, a responsible adult.

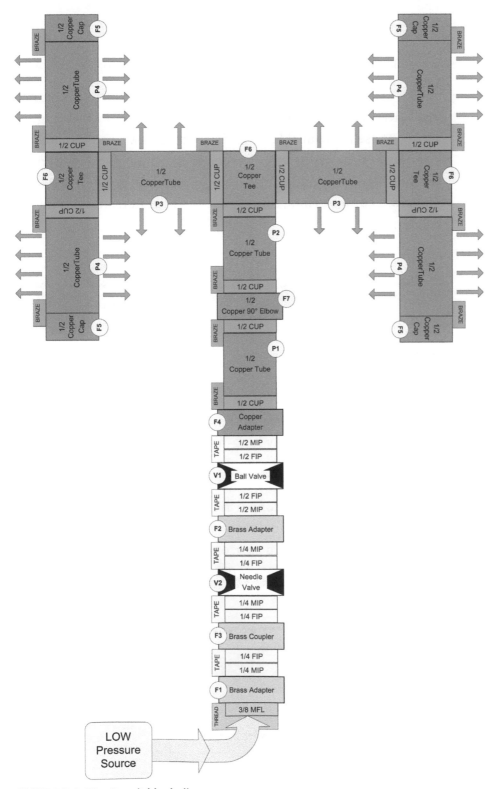

FIGURE 5-2: The fire pit block diagram

Tools

We will need a couple of additional tools for this project.

TUBE BENDING SPRING   We'll discuss different ways to bend tubing a little later on, but the tool we'll end up using is a bending spring. These are available at hardware stores as a set to match a variety of tube diameters. I've seen long bending springs and bending springs intended to go into, rather than around, tubes. Those are harder to find, so we'll use the common short spring bender. (See Figure 5-3.)

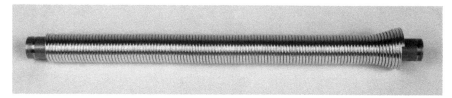

FIGURE 5-3: Tube bending spring

HEAVY WIRE CUTTERS   We'll be cutting some lightweight chain links. There are lots of ways to do this; I use heavy wire cutters. I particularly love my long-handled cutters. I bought them at my local discount import tool store and they get a surprising amount of use. The long arms provide so much leverage that they will easily cut copper rivets, small chain, and other things normally too difficult for wire cutters. You could use a hammer and cold chisel, a saw, or other methods to snip the links, but for a tool you'll end up loving, try one of these. (See Figure 5-4.)

FIGURE 5-4: Long-handled heavy wire cutters

Construction

Some readers may wonder why I'm not suggesting using propane-rated refrigeration copper coil. Since it's already bent into a circular coil, it would be easy to bend into a spiral, flare and cap each end, and add jets. We will, in fact, use refrigeration coil in another project. My intent with this book and these projects has more to do with skill acquisition than the projects themselves. You'll notice that, in order to provide breadth and exposure, I mix up approaches to problems from project to project. Coiled copper tube is a great alternative to brazing for this project.

If you work through Chapter 8, "The Venturi Burner," where we learn flaring, you may decide to go that route instead of our exercise in bending and brazing.

Since the fire pit is circular, that's the general shape we'll aim for with the new burner. This burner will be center-fed into four equal length curved arms. (See Figure 5-5.) We could bend a ring and feed into one side of it, but that wouldn't distribute the pressure as evenly. We could add secondary rings or feed lines inside the main circle, but I'll leave that to you as an upgrade.

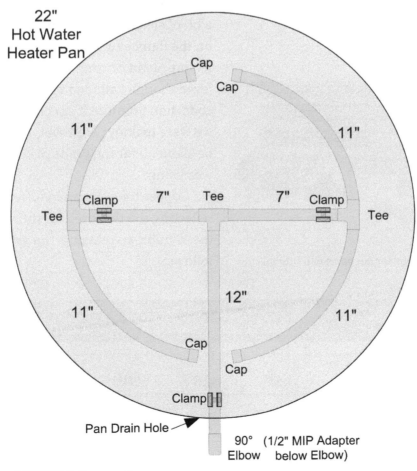

FIGURE 5-5: Burner diagram

1. Cut the Tube Lengths

Here, we return to our old friends, the tubing cutter and reamer (unless you went the hacksaw route). Cut the following lengths of tube:

1. (P1) One (1) 2″ (50 mm)
2. (P2) One (1) 12″ (300 mm)
3. (P3) Two (2) 7″ (180 mm)
4. (P4) Four (4) 11″ (280 mm)

Ream their ends so that there is no internal lip.

The set of copper burner parts should look like what you see in Figure 5-6.

We'll dry-fit the parts first to spot any anomalies before we start bending and brazing. (See Figure 5-7.)

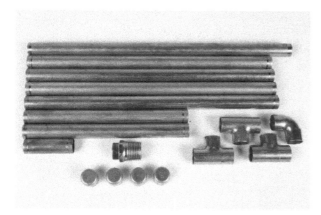

FIGURE 5-6: **Copper burner cut parts and fittings**

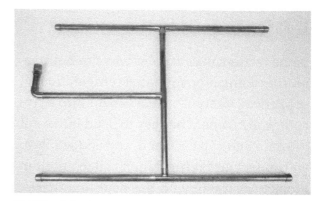

FIGURE 5-7: **Pre-bending, dry-fit the parts**

2. Bend the Four 11″ Segments

Because it impacts the success of the bends, I should come clean about something. Readers with a discerning eye may have noticed a bit of sleight-of-hand in the earlier chapter on the flambeau. I provided a lovely little lecture about brazing as a code requirement (even though I alluded to the fact that the code didn't require it specifically for the part we were making). However, I managed to be silent about the grade of copper tubing I was using.

Copper tube comes in different grades reflecting the thickness of the walls. For the ½″ tube we're using, the grades are as follows:

NOMINAL PIPE SIZE	OUTER DIAMETER	INNER DIAMETER (INCHES)			WALL THICKNESS (INCHES)		
		K	L	M	K	L	M
½″	⅝″	0.27	0.545	0.569	0.049	0.035	0.025

For use, *as piping,* in propane systems, the NFPA 58, Liquified Petroleum Gas Code and NFPA 54, National Fuel Gas Codes both state that copper tubing must be of type K or L as defined by:

- ASTM B 88, *Specification for Seamless Copper Water Tube* or;
- ASTM B 280, *Specification for Seamless Copper Water Tube for Air Conditioning and Refrigeration Field Service*

In this book, we are not using copper under pressure for piping; we will only use it at atmospheric pressure as a burner. We are relying on the heat resistance of brazing rather than its strength.

In the flambeau project, I wasn't specific about the grade of tubing I was using, but if you know the markings, you would have noticed in the pictures that the tube had red markings indicating it was type M (gasp!). I chose this because the thinner walls make it easier to file and pierce the jets (it also doesn't hurt that it's cheaper).

However, tight bends in type M are a nightmare. Figure 5-8 shows some examples of attempting the bend we want using type M tubing.

For this reason, I have specified type L for this project. You may notice in the pictures that it has blue markings. The thicker walls sustain themselves better during a tighter bend. The downside is that it takes a bit longer to cut with the tubing cutter and slightly

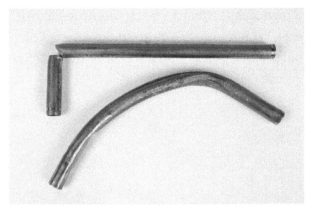

FIGURE 5-8: Failed bends in type M tubing

more filing to get deep enough that you can punch the jet with a wire nail.

For this project, it's not critical that we achieve perfect arcs with our bent tube segments. Our goal is to reasonably disperse propane into the lava rock of the fire pit. Perfect, even curves are a wonderful and heartwarming thing, but since our burner is going to be hidden beneath a bunch of rocks, ideally no one will ever see it. There is also a degree of randomness in the way the rocks disperse the vapor, which is part of the appeal.

One of the goals for this project is to learn a bending process that we can use again later when we want to make even tighter bends. There are many ways to bend copper tube. The ideal is to use an industrial CNC tube bender. However, since these cost hundreds of thousands to millions of dollars, we'll skip that one for now (but dream of it at night!). The most common method is probably the tube bender. This is a device, specific to a particular tube diameter, which allows you to make a series of successive bends in a section of tubing. (See Figure 5-9.)

FIGURE 5-9: **Tube bender**

The tube bender is not a favorite tool of mine. Far too frequently, the tube bender leaves a crimp, flat, or crease in the tubing. It can also be tricky for new users to handle. Professionals do amazing things with them, but I'm going to recommend an old-school method that, when done carefully, achieves consistent results without a lot of complexity.

The tool I'm recommending is the bending spring. The bending spring slips around the outside of the tube. You need a different bending spring for each size of tube you want to bend. Usually, you place the tube, covered with the spring, against your knee and pull the ends of the tube toward you. The spring protects the walls of the tube, helping it to avoid crimping, cracking, or folding. We'll be using a vise to hold the work, and pushing instead of pulling, but the principle is the same.

The bending spring is cheap and effective, but on a short piece of ½″ type L tube, such as the 11″ (280 mm) (P4) pieces we'll be bending, it is difficult to use as I just described. The bend is almost impossible for most people to make. There are ways around this—you could make the bend with extensions to the sides for additional leverage or some other scheme. But we're going to use the spring in combination with two other techniques to get a great result without straining ourselves.

We will use a bending approach by filling the tube with dry sand and heating it with a torch. When it's almost glowing hot, we'll slip the spring over and bend it with very little effort. Any of these techniques—spring, sand, or heat—can be used by themselves, but combining them makes everything easier (at least in my shop).

Another goal is to end up with braze-able ends. This is an issue because, in the course of clamping and bending the tube, a significant amount of uneven force will be applied to the ends of the tube. If we mar or warp them, we'll compromise our ability to braze a clean joint later. A lot of bending techniques abuse the ends. If we'd bent the entire circle and cut sections out of it, the ends wouldn't be quite square and we could have difficulty in a fitting.

We'll get around this problem by using a copper tee in a vise as a jig to hold the tube for bending. You don't need to clamp the tee so tight that it deforms. Clamp it just tight enough to keep it from moving. The tee will take a little abuse, so you may want to buy an extra one for this purpose. (See Figure 5-10.)

FIGURE 5-10: Mount the workpiece in a tee in a vise.

Plug one end of the tube you're going to bend with a wad of paper. You shouldn't need tape or anything else to hold it; the paper is only there to temporarily keep the sand in. Place the plugged end into the tee in the vise. Pour clean, dry sand into all but the last ½" (10 mm). I'm using a plastic funnel, but you can make one out of paper. (See Figure 5-11.)

Plug up the top with another wad of paper. Try to get all the paper into the tube so that there aren't any pieces hanging over the side that the torch will ignite.

Put on your gloves and safety glasses. Ignite your torch and begin heating the tube. Play the flame up and down the tube, avoiding the last inch on either end. Heat all sides of the tube until you notice a color change. You don't have to get the tube glowing hot, but you want to get close to it. (See Figure 5-12.)

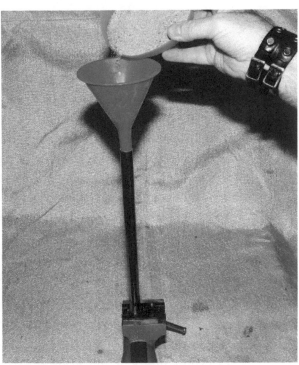

FIGURE 5-11: Fill the tube with sand.

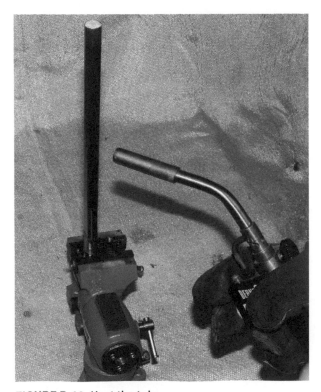

FIGURE 5-12: Heat the tube.

Safety Note

This project requires handling an extremely hot section of tubing. You must wear appropriate gloves. Ideally, use heavy welding gloves. If you must use gloves lighter than welding gloves, use heavy kitchen mitts over them. You may scald them, so don't use ones that someone (perhaps your significant other) cares about. Even welding gloves will only give you seconds of safety while handling metals at high heat. You can get a nasty burn through welding gloves in considerably less than 20 seconds if the material is hot enough. If you start to feel the heat through your gloves, you have only a couple of seconds to get your hand out of the glove before you can get burned. Think before you touch. Be smart; use gloves in conjunction with pliers or other mechanical holding tools and be aware of where things will fall if you have to drop them in a hurry. High heat and open-toed footwear are incompatible.

Pro tip: The spring can often be difficult to remove. If the tip of the narrow end of your spring isn't angled out, it can dig into the tube and make removal difficult. Before using it, take some pliers or vise grips and try to bend the spring's tip a tiny bit outward. Even that slight bend will help immensely.

As soon as you finish heating, remove the flame and slide the spring down all the way down the tube, grab the top end with your gloves, and bend the tube away from you. (*Please* read the safety note regarding appropriate gloves before you grab the nearly glowing hot metal.) This will create about half the bend we want. (See Figure 5-13.)

Twist the spring so that it opens up and pull so that it slides off the tube. If the tube comes out of the tee/vise while removing the spring, place it back into the tee and try again. Twisting to slightly unwind the spring and pulling upward should allow the spring to slide up and off the tube. Be gentle with

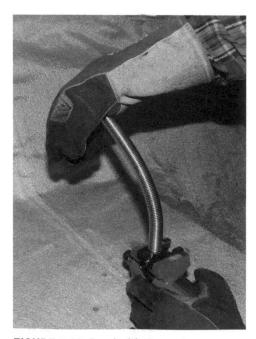

FIGURE 5-13: Bend with the spring.

your motions; you don't want the paper plugs to fall out. The sand inside is extremely hot! (See Figure 5-14.)

FIGURE 5-14: Twist and pull the spring off.

Using pliers or vise grips, pull the tube out of the tee/vise, reverse the tube, and put the other end back into the tee. The tube is too hot to do this with your gloves unless you wait for it to cool down.

Heat the bottom half of the tube with the torch so that it gets back to the bending temperature. When it's ready, put a little weight on the end of the tube to complete the full bend. (See Figure 5-15.)

Lift the tube out of the tee/vise with some pliers or vise grips, position it over the sand receptacle, and use something to pull the paper plug out (needle-nose pliers, a needle, wire, or similar). (See Figure 5-16 on the next page.) Be careful! It's worth saying again: the sand is extremely hot! The tube will cool down much quicker once the sand is out. If you're not in a hurry, you can wait and take the sand out once everything cools.

As noted earlier, we don't need to achieve perfection with these bends (unless it satisfies some deep internal need). If you want to continue to improve the bends, keep the sand in the tube and continue the bending process (you don't need the springs for subsequent small bends) until you like the bend you get.

FIGURE 5-15: Complete the final bend without the spring.

FIGURE 5-16: **Pour out the hot sand.**

Once you have all four of the 11″ (280 mm) pieces bent, dry-fit all the parts to see how it looks. (See Figure 5-17.)

Make any modifications you determine are needed, then we'll move on to brazing.

FIGURE 5-17: **Dry-fit the bent parts.**

3. Braze the Burner

For details on the brazing process, refer back to Chapter 4, "The Flambeau." You can perform the 16 required brazes in any order that pleases you, but here's the order I performed them:

1. The caps (F5) on the ends of all the curved arms. (See Figure 5-18.)

FIGURE 5-18: **Braze the caps.**

2. The tees (F6) on the ends of the two 7″ (180 mm) segments and one 12″ (300 mm) segment. (See Figure 5-19.)

FIGURE 5-19: **Braze the tees.**

3. The 90° elbow (F7) on the opposite end of the 12″ (300 mm) segment. The plan is for the stack of fittings we used on the flambeau to route downward, so make sure to rotate the elbow a quarter turn from the plane the tee is in. (See Figure 5-20.)

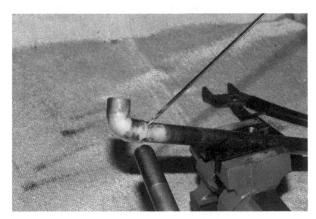

FIGURE 5-20: **The 90° elbow faces downward**

4. The 11″ arms (P4) into the tees on the two 7″ (180 mm) segments. We want the burner to be able to lay flat, so we want the arms and the segment all in the same plane. (See Figure 5-21.)

FIGURE 5-21: **Braze the bent arms into the tee.**

5. The 7″ (180 mm) arms into the tee on the 12″ (300 mm) feeder arm. Again, we want these to be able to lay flat, so be careful about the angles. (See Figure 5-22.)

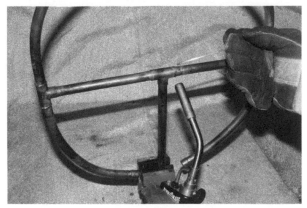

FIGURE 5-22: **Braze the side arms into the feeder tee.**

6. The ½″ MIP adapter (F4) onto the 2″ (50 mm) segment (P1). (See Figure 5-23.)

FIGURE 5-23: **Braze the threaded adapter to the short arm.**

7. The 2″ (50 mm) segment into the 90° elbow (F7). (See Figure 5-24.)

FIGURE 5-24: **Braze the short arm into the 90° elbow.**

This is all the brazing for the project. With brazing complete, the burner should look something like what's shown in Figure 5-25.

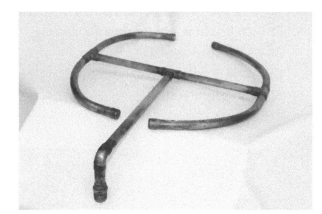

FIGURE 5-25: **The brazed burner**

4. Punch the Jets

The number of jets is arbitrary. We can get by with fewer than the flambeau, since the lava rock we'll use will help us disperse the propane. I'm using a total of 48. (See Figure 5-26.) You can use however many you want. If you want to start with just a few, it's easy to add more later. Even random placement has some merits since we want a "natural" distribution of flames. Personally, I went for maximum coverage.

I positioned the jets closer to the side than the top of the burner so that the propane will disperse among the lava rock and not shoot straight up.

5. Punch Drain Holes in the Pan

Since we don't want the pan to fill up with water if it rains, we need to poke a few small holes around the inside bottom edge and a few across the bottom. Put a piece of wood under the pan and use a small nail or a wood screw and a hammer to punch six to eight small holes around the bottom rim and five to six on the base. (See Figure 5-27.) Try to keep them as low as possible on the rim so they'll drain the bottom of the pan. It's undesirable to have too many holes, since propane sinks instead of rising, and we don't want all the propane escaping out the bottom of the pan before it lights.

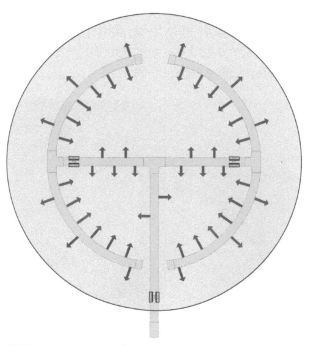

FIGURE 5-26: **Jet positions**

FIGURE 5-27: Punch some drain holes.

6. Mount the Burner in the Pan

Lay the completed burner in the empty pan (A1) with the elbow and ½″ adapter sticking out. Position it so that you can get the best dispersion from the jets. The dispersion pattern is up to you. You can go for a random pattern or something with an even distribution of flames. We'll use three of the conduit hangers (S2) to hold the straight arms of the burner to the pan. The mounting is not critical, but I positioned mine 6″ (150 mm) from the center on the short arms and 10″ (250 mm) on the long arm.

Remove the tightening screw from the hangers, open the jaws, and slip it over the burner at each of the target locations. Set the hangers, base side down, on the pan with the burner resting in them. With a marker or pencil, mark the location of where the hangers sit. We'll want a mark where the center screw will go, but that's hard to get at with the burner in the way, so you may want to outline the base of the hanger, remove the burner, reposition the hanger in the outline, and mark the center hole.

Drill a ¼″ hole through the pan at each hanger's center hole mark. The aluminum of the pan is fairly thin, so you will get a cleaner hole if you set the pan on a piece of wood and drill through the pan into it.

Use a ¼″ × ½″ bolt, lock washer, and nut to fasten each hanger in position (S3, S4, S5). (See Figure 5-28.)

FIGURE 5-28: Hangers mounted in the pan

As an optional step you can add a drain screen (A3) (see Figure 5-29) to keep material from falling out the pan's drain hole. Just cut a hole in the screen's center and slide it over the copper burner. It's not strictly needed if you're using lava rocks, but for pea gravel it will help a lot, and it will help some with sand (see the sidebar, "Fire Pit Filler Material.").

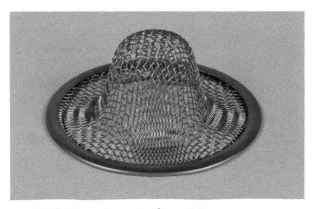

FIGURE 5-29: Drain screen for a sink

Fire Pit Filler Materials

The biggest danger with material used for a fire pit is that something will explode and shoot hot pieces of material at people. This is potentially due to water trapped in a porous material, heating until it turns to steam and causing an explosion. Alternately, some materials have internal structures or flaws that can respond to excessive heat catastrophically. A few materials are designed for use in high heat and we've chosen the easiest to acquire: BBQ grill lava rock.

Lava rock is just what it sounds like: rock produced from the ejection of lava from a volcano. Most lava rock used for barbeque is scoria, a rock that solidifies as gas is erupting, forming an almost foamy matrix (technically called *highly vesicular*). This is the rock associated with cinder cones. Being born in fire, it remains able to withstand high heat without problems.

A common filler in many Internet fire pit projects is sand. Sand is appealing in that you can do a kind of "Zen fire garden" thing with it. The downside to sand is that it can retain moisture. Trying to light the fire pit if the sand is wet will be frustrating, and the wet sand can clog up jets. Sand also does not disperse gas as easily as rocks, and it could potentially create a back-pressure situation—which normally isn't a problem, but if you undertake the enhancement option to add an in-line air mixer, back pressure would be unsafe. Some sources also suggest that sand could present micro-explosive risks when heated, but this is unlikely with clean, dry silica sand. Wet sand, on the other hand, could potentially jet steam under some circumstances. Even more likely is that wet sand would create a pool of propane under it that results in a fireball when lit. If you choose to use sand, use a cover and keep it dry.

Pea gravel is another option. Unfortunately, you can find sources that state it is absolutely safe and sources that say the opposite. Many people will choose to use a layer of pea gravel beneath the burner and lava rock, or our next contender, fire pit glass.

Arguably, the very best fill material is special fireplace or fire pit broken glass made especially for this purpose. It comes in beautiful colors and is safe to heat in the pit. It's been tempered to make sure it doesn't break easily and tumbled to take off any sharp edges. The tumbling also serves the purpose of enhancing the ability of the propane to disperse the gas and not leave sharp edges or points that get extra hot. Just make sure you're buying glass made for this purpose.

Put the adapter end of the burner out through the hole in the pan and place the burner into the hangers. If you used a screen, be sure that the mesh is able to cover the hole and isn't blocked by a hanger. Replace the tightening screws on the hangers and tighten to hold the burner in place. (See Figure 5-30.)

FIGURE 5-30: Burner mounted in the pan

7. Prepare the Stand

Initially, I struggled with what kind of stand to recommend. The concern is that the pan will eventually hold around 21 lbs (9.5 kg) of lava rock, which will be extremely hot after an evening of using the fire pit. We need a stable base that will not allow the pan to easily fall over if something knocks into it. The desired height is also something that will vary from person to person. I wanted mine to be about 18–24″ (45–60 cm) off the ground.

Often, in moments of doubt like this, I ask myself, what would Buckminster Fuller do? And, as is so often the case, the answer is, dynamic tension! Specifically, in our case,

it would be . . . tensegrity! (Please imagine me shouting this in a deep voice with my finger pointing dramatically at the sky.) *Tensegrity* refers to tension with integrity, and generally uses isolated rigid members held together with tensioned wires, rope, or chains as a means of achieving structural stability. Truly awe-inspiring structures can result from this principle. These structures are surprisingly lightweight, while providing tremendous strength and rigidity. (See Figure 5-31.)

FIGURE 5-31: Tensegrity structure—Science Park, Science City, Kolkata PUBLIC DOMAIN; ORIGINAL PHOTO BY BISWARUP GANGULY.

Our needs are modest, so we'll use the simplest tensegric structure. Technically referred to as a *three-prism*, sometimes called a *tensegrity tripod*, the design is simple, strong, and stable. There's an unlimited number of other stands you can build, but this one has an interesting hypermodern feel and has the advantage of being portable. Building tensegrity structures is a mania for me and many people I know; be careful, you may get hooked!

I'm going to specify the stand in Imperial units and offer what I hope will be common metric equivalents. The hole sizes are not critical as long as the quick links you buy will easily fit in them, so use the parts you can acquire rather than going out of your way if I specify an uncommon size.

We'll use a 10′ (3 m) section of 1″ (25 mm) EMT conduit (S1) cut into 3 equal sections for the legs. You can cut the EMT with your tubing cutter, but a hacksaw will be quicker. Many people frown on using tubing cutters for conduit, but this is because of the risk of the sharp interior lip cutting into the wires that the conduit will carry. Since we're using the conduit off-purpose (with no wires), that's not an issue. For this use, don't worry about reaming.

We'll drill a ¼″ (6 mm) hole ³/₁₆″ of an inch (5 mm) from each end. We want the holes to be aligned on the tube, so look for the faintly evident seam on the EMT and use it for registration. Drilling on metal pipe can be frustrating since the drill bit has a habit of wandering. We'll use a quick three-step process to keep this from happening.

Place a small piece of tape near the end of the leg leaving a bit of the seam showing between the tape and the end of the EMT. Use a ruler, lined up along the seam, to mark a lateral line along the seam and a perpendicular line at ³/₁₆″ (5 mm) from the end of the tube. Use the center punch to create a divot through the tape into the metal. (See Figure 5-32.)

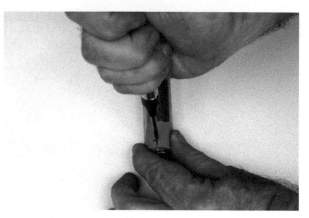

FIGURE 5-32: Mark the hole location on the EMT.

Use a very small drill bit; size isn't critical—something near ¹/₁₆″ (2 mm) to drill a pilot hole through the divot made by the center punch. (See Figure 5-33.)

Finally, switch to the step drill bit and drill a ¼″ (6 mm) hole. (See Figure 5-34.)

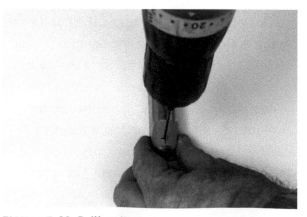

FIGURE 5-33: Drill a pilot hole.

FIGURE 5-34: Drill the final hole with the step drill bit.

We'll also use approximately 20′ (6 m) of #3 double loop chain (C). That's a lightweight chain that should be relatively easy to cut. (See Figure 5-35.)

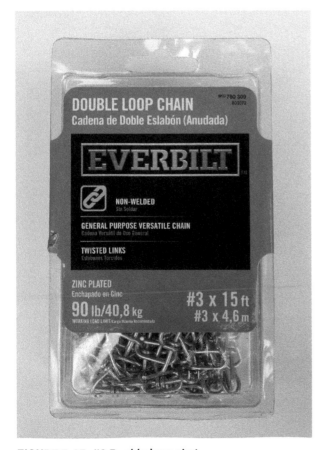

DOUBLE LOOP CHAIN
Cadena de Doble Eslabón (Anudada)

SKU 760 309
803072

EVERBILT™

NON-WELDED
Sin Soldar

GENERAL PURPOSE VERSATILE CHAIN
Cadena Versátil de Uso General

TWISTED LINKS
Eslabones Torcidos

ZINC PLATED
Enchapado en Cinc

90 lb/40,8 kg
WORKING LOAD LIMIT/Carga Máxima Recomendada

#3 x 15 ft
#3 x 4,6 m

FIGURE 5-35: #3 Double loop chain

The ratios of the lengths of chain are more important than the specific length. Different brands of #3 double loop chain have slightly different lengths for the same number of links. You'll need to use chain that is the same length for all links. You'll also notice that the package of chain shown in Figure 5-35 is 15′ and I specified 20′. You could buy an additional section, but you'll need to buy 6′ more, not 5′ (since the package will end up leaving you an odd unusable length). Otherwise, you can have the store cut you a 20′ length. (I'm not sure how double loop chain is sold outside the United States, so please use the link lengths below to determine how much to buy.)

I'm going to describe the lengths of chain using the number of links rather than length. This will allow it to work with whatever brand you end up with.

Cut the chain into the following lengths:

- (C1) Three sections of 29 links
- (C2) Three sections of 25 links
- (C3) Three sections of 12 links

Open up the quick links (S6) and put the end through each of the holes in all tubes. Do this by putting the threaded end into the tube and out through the hole. Then rotate the link so that the opening is at the top. You can see how this should look in Figures 5-37 and 5-38 on the next page.

You will find that the links want to fall out of the tube, or the chain wants to fall off the links. Get into the habit of closing the lock on the quick link whenever you're not working on it. When we tension the chains it will be critical that the links are closed. Quick links have relatively little strength when open. Figure 5-36 shows what your links will look like if you tension without closing them.

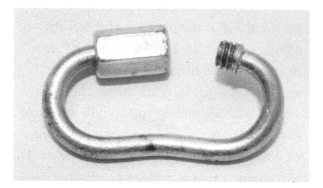

FIGURE 5-36: Bent quick link

Each piece of conduit, or leg will get the same set of parts. The "bottom end" will get the eye end of the turnbuckle (S7) and a 25-link section of chain (C2). (See Figure 5-37.) Make sure the turnbuckle is in the fully extended position—in other words, the hook and eye are as far apart as possible. If necessary, you can twist one end or the other to get both of them evenly extended.

The "top end" will get a section of 12 links (C3) and a section of 29 links (C1). (See Figure 5-38.)

FIGURE 5-37: Bottom parts on each leg

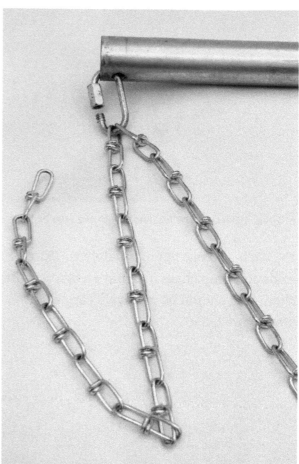

FIGURE 5-38: Top parts on each leg

8. Assemble the Stand

While this stand is lightweight and portable, it can be extraordinarily frustrating to assemble unless you follow these instructions carefully. There are hundreds of slightly wrong ways to assemble this that will frustrate you to the point of madness. (Or at least I tried hundreds of wrong ways to do it before I got a working assembly pattern. The degree of madness I incurred is a matter of lively debate in my household.)

8a. Lay Out the Legs

Arrange the legs in a triangle with the tops extending out beyond the vertices. (See Figure 5-39.)

FIGURE 5-39: Step 8a, laying out the legs

When attaching chains for this stand, you will always keep the order of the items on the quick link the same:

1. The original 25- (bottom) or 29-link (top) chain

2. The turnbuckle (bottom) or 12-link chain (top)

3. The 25- (bottom) or 29-link (top) chain from the next leg

There should never be a need to remove the first or second of these. The only thing we'll connect to the quick links during assembly is the third. This also means that, if you take the stand apart for transport, *do not* remove the first or second from the above list from a quick link.

8b. Attach the Bottom Chains

Connect each of the 25-link segments to the bottom of the next leg in clockwise order. Make sure the quick link is angled toward the leg connecting to it. (See Figure 5-40.)

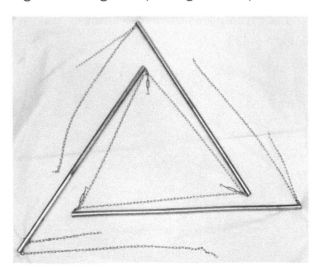

FIGURE 5-40: Step 8b, connecting the bottom chains

8c. Overlap the Legs

One of the many ways that assembly can go wrong is if the legs cross each other in the wrong order. The correct order is so that each leg goes over the clockwise leg and under the counter-clockwise leg. (The direction of clockwise versus counter-clockwise is arbitrary, but each leg must go over one and under another.) (See Figure 5-41.)

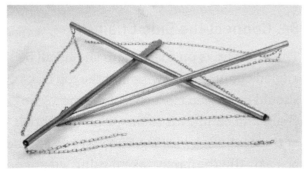

FIGURE 5-41: Step 8c, overlapping the legs

8d. Connect the Top Chains

This is easier with a helper, but it's possible to do alone. Connect a 29-link chain to the top of the next counter-clockwise leg (the opposite direction of the bottom chains. The chain will go *under* the leg and, again, the quick link will angle toward the leg connecting to it. Look carefully at Figure 5-42 to understand.

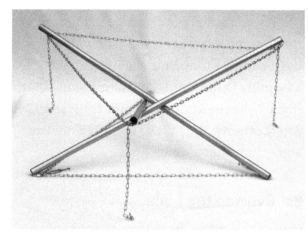

FIGURE 5-42: **Step 8d, connecting the top chains**

As you connect the top chains, the tops of the legs will rise and rotate. As the picture shows, the structure should be able to stand once all three are connected. It's not stable until we connect the side chains, so don't get too excited yet!

8e. Connect the Side Chains

The quick links on the bottom should all be angled counter-clockwise and the top links should be angled clockwise. Take the 12-link chain and connect it to the turnbuckle hook of the next clockwise leg. The first two should be easy; the third may take putting weight on the top of the leg to get it connected.

Place the pan and burner into the top triangle. It will require spreading the chains a small amount. The bottom of the pan should rest on the conduit legs and the chain should be just below the top lip of the pan on each side.

The pan will flex a little as the chains compress it on three sides. We could have made the top triangle larger so this wouldn't happen, but without increasing the length of the legs (requiring more conduit) or decreasing the size of the bottom (reducing the size of the base and related stability), the stand would have been lower and the fitting stack would be too long. A small amount of structurally insignificant warping in the pan was the design compromise I made. Too much warping would have created a catastrophic failure in the pan, but we'll avoid pushing it that far. We'll also get some compression on the sides of the pan to hold it in place (a classic tensegrity solution!).

Once all three side chains are connected and the pan is in place, tighten each of the turnbuckles the same amount. (See Figure 5-43.)

FIGURE 5-43: **Step 8e, tightening the turnbuckles**

Start with tightening them a quarter of the way and keep going until the structure feels strong and the chains have no slack. Tighten evenly in small amounts; don't tighten one turnbuckle too much before tightening the others to match. There may still be a small amount of compression between the top and bottom triangles, but as long as the pan is resting on the legs, this won't matter.

If you remove the pan, the stand is an attractive structure all by itself! (See Figure 5-44.)

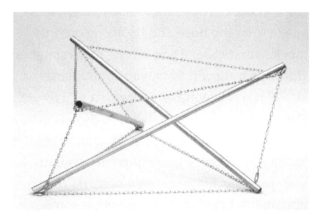

FIGURE 5-44: Tensegrity!

8f. Mount the Pan

With the pan inside the top chains, we're going to add another safety feature and attach it so that it can't come out. At the fifteenth link on each side of the top chain (the center link), we're going to drill a ¼" hole through the pan and use a ¼" × ½" bolt, lock washer, and nut (S3, S4, S5) to attach the chains to the pan. Mark where the link is on the pan and drill a ¼" hole ½" down from the link. Depending on the brand of chain you purchased, you may need to open up the link

by twisting a flat head screwdriver in it to open it up and let the bolt pass through (or use smaller bolts as long as the head won't pass through the link). Don't open the link any more than necessary; you don't want to break the link—just make the passageway slightly larger.

Push the bolt through from the outside. You may need to push down some of the sharp edges on the inside of the pan, but be careful, it's sharp. Put on a lock washer and nut and tighten. (See Figure 5-45.)

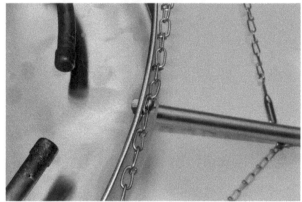

FIGURE 5-45: Step 8f, attaching the pan to the chains

9. Attach the Rest of the Fitting Stack

Tape the end of the copper ½" MIP (F4) and thread the ball valve (V1) onto it. Hold the copper fitting in place with a wrench while tightening the ball valve. You'll want to be careful to both get it tight and position the ball valve facing out for easy access.

Continue with the rest of the parts in the stack in the same manner as on the flambeau:

1. (F2) ½" MIP × ¼" MIP brass adapter

2. (V2) ¼″ MIP × ¼″ FIP needle valve

3. (F3) ¼″ FIP × ¼″ FIP brass coupler

4. (F1) ¼″ MIP × ⅜″ MFL adapter

Assembly is complete! Hopefully your portable fire pit looks something like the one shown in Figure 5-46.

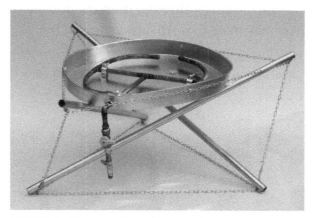

FIGURE 5-46: **Assembly steps completed**

10. Filler Material and Testing

Take your lava rocks (A2; assuming you made the same filler choice I did) and arrange them over the burner. Distribute them so that they evenly cover the whole pan and cover all of the copper. (See Figure 5-47.)

FIGURE 5-47: **Fire pit with filler**

Part of me really hopes that a little voice inside your head just said, "Hey! He didn't leak-test the burner before he covered it!" It's not that I want to encourage voices in your head, but if you do have one that reminds you to do things like leak-test, you could do worse. In this case, if there is a small leak in the burner under the lava rock, it's serving as a jet, so we're okay with it! Have no fear, we'll leak-test the parts outside the pan before we light up.

Thread the hose from the low-pressure source onto the flare fitting on the bottom of the stack (F1). There should be just enough space to fit the hose and its fitting onto the stack. Worst case, if something has happened that won't allow it to fit, you can replace the ¼″ FIP × ¼″ FIP coupler (F3) with a ¼″ FIP × ¼″ FIP 90° elbow.

Consider the area in which your hose is resting. You don't *ever* want people walking on your precious high-pressure hose. If I'm in a situation where I'm concerned that someone might step on my hoses, I will route them through a section of PVC pipe or EMT conduit. This is only serving as a protective shell, not actual pressurized piping. Any diameter that the female flare fitting at the end will fit easily through will work. Unless you've it encased in a hard shell like this, do not ever cover hose with a carpet, mat, or other material that will let people step on it without knowing.

Check that all four valves; cylinder valve, source cutoff, needle valve and effect cutoff are closed. Get your leak-testing kit ready to check for leaks. In all cases, if a leak is found, shut off all valves and tighten the offending joint. Then start the leak-testing sequence over again.

Test for leaks by spraying with the soapy water leak-testing solution. Respond as described above if any are found.

1. Close the fire pit ball valve.

2. Open the cylinder valve all the way then back a quarter turn.

3. Test the section between the cylinder and the fire pit ball valve.

Hopefully everything worked great. After passing the leak test, you're ready to hang out with friends and enjoy your portable fire pit!

Operation and Shutdown

To use the fire pit follow this sequence:

1. Close the needle valve.

2. Open the cylinder valve all the way, then back a quarter turn.

3. Open the source cutoff ball valve.

4. Open the effect cutoff ball valve.

5. Light your fireplace lighter and hold the lit flame directly over the nearest arm of the burner.

6. Open the needle valve all the way while waving the lighter back and forth, directly over the lava rocks.

7. When the flames ignite, turn down the needle valve to the desired level. You may want to use it all the way open.

The shutdown procedure is as follows:

1. Close the cylinder valve all the way.

2. Close the source cutoff ball valve.

3. Close the needle valve all the way.

4. Close the effect cutoff ball valve.

5. Allow the fire pit to cool undisturbed until it's no longer hot. This may be as long as 45 minutes if it was in use for a significant amount of time.

Safety Note

If the flames don't ignite within 20 seconds, shut down the effect cutoff valve, step away, and wait for 5–10 minutes to let the propane clear before trying again. To put it mildly, it is undesirable to have a volume of unlit gas build up and create an eyebrow burning fireball (or worse). Seriously, this is the biggest danger for this project. Propane doesn't rise, so it can pool in the pan and build up a nasty surprise. You may even want to remove the lava rocks from one jet and light it directly if this is a problem.

Don't underestimate how hot the rocks and pan will get. Do not touch them directly while they're lit or until they've had considerable time to cool down. If you must fuss with the rocks while the pan is lit (and I understand that need), do so with a piece of metal. Personally, I use a handy tire iron, but there are lots of good alternatives. Just don't use bare hands because you think there aren't flames on that part.

Enhancements

The most straightforward change is to make a different stand. A welded stand, something like a potted plant holder, could be made out of rebar for a fun weekend project. You could even braze a stand out of copper tubing. Just make sure to fully support the pan so that it can't be jostled easily. Pay attention to how stable your stand is so that accidents are avoided. You could also play with the proportions of the tensegrity prism to create a higher or broader stand. Conduit is frequently sold in 5′ sections that would allow you to buy three long, pre-cut legs to make a table-height pan.

An enhancement I've adopted is to use pea gravel as the bottom layer in the pan, with the pan filled up to just below the jets. This serves to provide more coverage from the lava rocks without heating the pea gravel directly in a flame. Since the pan is easy to dump out, you could have different materials to try in different circumstances. (See Figure 5-48.) (Consider storing different filler materials in 5-gallon buckets with either covers to keep the rain out or holes punched in the bottom to allow drainage.)

Another simple enhancement, if you don't want to use the exact fitting stack from the flambeau, is the idea mentioned above to replace the ¼″ FIP × ¼″ FIP union with a ¼″ FIP × ¼″ FIP

FIGURE 5-48: Pea gravel bottom layer

90° elbow. This provides extra clearance and generally makes hooking the hose up easier. I debated making this a standard part of the design for this project since it's a significant improvement, but I wanted to keep the parts cost as low as possible and reuse whatever we could from the previous project. (See Figure 5-49.)

Well-made commercial fire pits will add an air mixer to the plumbing to achieve a complete stoichiometric (say it with me, *stoy-key-oh-metric*) burn from the propane. Since this has to mix the appropriate amount of air with the volume of propane being used, this can vary based on pressure, plumbing diameters, and other factors. While it is possible to build a Venturi-based, air injection system (we'll build one in the chapter on the foundry burner), commercial units are worth exploring for a use like this.

We mentioned using higher pressure. Before I'd go above 1 psi (remember, we're at about ½ psi now), I'd use a steel pan and consider how I wanted to use this fire pit. With great big flames at knee level, the danger to self and others increases significantly. This isn't an enhancement I'd recommend with this design.

It's also possible to add a manual or electric ignition to the fire pit. Kits are sold for either a piezo manual spark ignition or a battery-powered spark. These are relatively easy to include. Chapter 10, "The Pilot," will also provide you with a possible solution using a hot surface igniter.

FIGURE 5-49: 90° elbow

The Experimenter's Rubens Tube

6

IN 1904, HEINRICH RUBENS was a physicist well into a career studying waves. He was fascinated by every kind of wave he could find: electrical, acoustical, optical, or mechanical. A contemporary of Max Planck, Lord Rayleigh, and Marie Curie, Rubens made significant contributions to our understanding of subjects like blackbody radiation and other precursors to quantum theory. But it's an experimental device he constructed that made his name famous outside the world of physics. The device he built to visually demonstrate standing waves has become known to experimenters and makers around the world as a Rubens tube.

While Rubens initially used bells and tuning forks to create acoustic waves for his tube, modern Rubens tube users typically use a speaker to generate the desired sounds. These sounds are channeled into the end of a tube that is provided with a steady supply of gas (Rubens used coal gas, but you've probably figured out we're going to use propane). The propane exits via a series of holes and is ignited to produce a row of flames.

YouTube is filled with videos of Rubens tubes jumping to electronic dance music, Bach, and AC/DC. The resulting patterns are primarily stroboscopic; the speaker is acting as a vapor pump. While entertaining, this ignores the beauty and elegance demonstrated when the tube acts as a finely tuned instrument and sets up a visual demonstration of a standing wave.

Most music has few singular pure tones that will set up standing waves. The broadband energy, especially from the bass, tends to pressurize the entire tube evenly, shooting the entire row of flames up or down. The fascinating result of pure tones is the ability to create different pressures at different points in the tube. A gradient that stands still, low pressure and high pressure in the same space, creating elegant curves that we can sample at intervals, using flame height to measure the pressure at each point. The more you think about what's going on, the more interesting everything gets.

The theory goes something like this: if you play a tone through a speaker, and then into a tube, the moving speaker creates a wave, alternately compressing and rarefying the air (the other way to describe this is that it "makes sound"). (See Figure 6-1.)

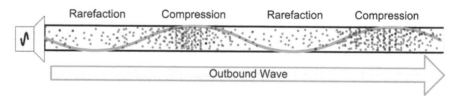

FIGURE 6-1: A wave in an open-ended tube

If you put a cap on the other end of the tube, the outgoing wave reflects back into the tube. The outgoing and reflecting waves interact. If the wavelength is just right, the outgoing and reflected waves line up so that the peaks (compression) of one wave match the troughs (rarefaction) of the other. This creates what is known as a standing wave.

Standing waves are just that, a wave that stands still. The bow wave on a boat, a vibrating cord or string, even rows of wave clouds in the sky are all examples of standing waves. The location where the two interacting waves are at the same level is called a *node*, which is also where the resulting amplitude is the highest. The location where the two waves cancel each other out is called an antinode, and the point where the amplitude is lowest.

The tube can produce standing waves when its length is a multiple of half the wavelength of the signal fed into it. The simple explanation is that, at each half wavelength, the wave is crossing the zero line. If you bounce the wave off the end of the tube when it's at zero, the wave starts over fresh going the opposite direction. If it bounces at any other point in the cycle, the reflected wave isn't equal (and opposite) to the outgoing wave.

As an example, the 66″ tube we're using works well with tones that have a wavelength very close to 22 inches (462 Hz). The 66″ tube is a clean multiple of half of that wavelength (11″). If you want to look for other candidates for standing waves, frequency is calculated as the velocity divided by the wavelength. The velocity of sound in propane is 10,157.48 inches/sec (258 m/s). (See Figure 6-2.)

As a result, when Rubens tubes are fed a pure tone (a sine wave) of an appropriate wavelength, they produce a beautiful example of a standing wave with a series of flames of different

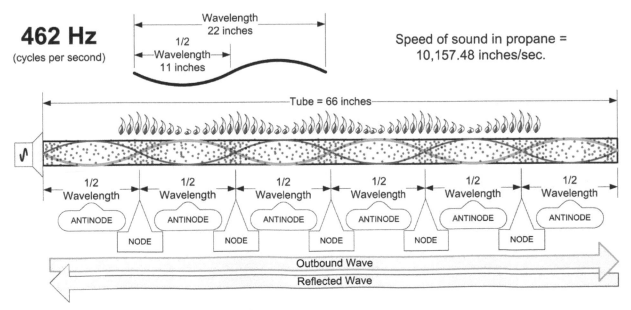

FIGURE 6-2: A standing wave in a Rubens tube

heights. From a different perspective, it's also a lovely example of how digital sampling works. While the analog wave in the tube is a smooth gradient, we are sampling the wave at discreet points with our jets, turning it into a series of flame height measurements. Rubens used this instrument as a scientific tool much as we might use an analog-to-digital converter today.

To tease you with a little more technical detail, the behavior we just described is true for low amplitude, or "not loud," signals. When louder signals are used, the behavior of the flames, *vis-à-vis* the nodes and antinodes, actually reverses! Since this book is focused on building projects rather than serving as a fluid dynamics primer, I'll let you dig into the wealth of excellent Rubens tube information on the Internet to understand why. Check out the list of links in Appendix B for good resources.

I initially intended to be very specific about how to put the parts together to build a Rubens tube, but as I played with mine, I found that I wanted to try a lot of different things. Rather than suggesting that I've found the perfect set of build parameters, I've made this project an "experimenter" Rubens tube. By this, I mean that we will construct this in a manner that will allow you to experiment with different approaches to tune and enhance it. This design works great as described below, but I have faith that you have even more interesting ideas that I haven't thought of.

Parts

The Rubens tube schematic should look very familiar after the last two projects, but I promised one for each project, so here you go (see Figure 6-3).

FIGURE 6-3: The Rubens tube schematic

Tube parts:

REF	ITEM	QTY
A1	Speaker system with (powered) subwoofer	1
C1	Plastic canvas	1
C2	Twist ties (4″ lengths of light wire)	14
C3	Newsprint (roughly 1″ × 4″ strips)	~100
C4	Papier-mâché solution (school glue and water)	12 oz (350ml)
M1	Nitrile glove (5 mil or thicker)	1
M2	Hose clamp 2½″	1
P1	Terminal fence post 2⅜″ × 66″	1
P2	Post cap 2⅜″	1
F1	Brass coupler ¼″ MIP × ¼″ MIP	1
V1	Needle valve ¼″ MIP × ¼″ FIP	1
V2	Gas-rated ball valve ¼″ FIP × ¼″ FIP	1
F3	Brass adapter ¼″ MIP × ⅜″ MFL	1
M3	Adhesive insulating foam, 1″ wide	1
M4	Vinyl electrical tape	1

Stand parts:

REF	ITEM	QTY
S1	2″ × 4″ × 12″	2
S2	Adjustable wood adapter 2⅜″	2
S3	Adjustable leveling feet (pack of 4)	1

The block diagram should be pretty familiar, as well. (See Figure 6-4 on the next page.)

With apologies, I'm going to use inches as the main unit of measurement in this and subsequent projects. I'll add metric measurements where it's not related to parts I've specified in Imperial units. If you need to change out parts for locally available metric equivalents, please change other necessary measurements appropriately.

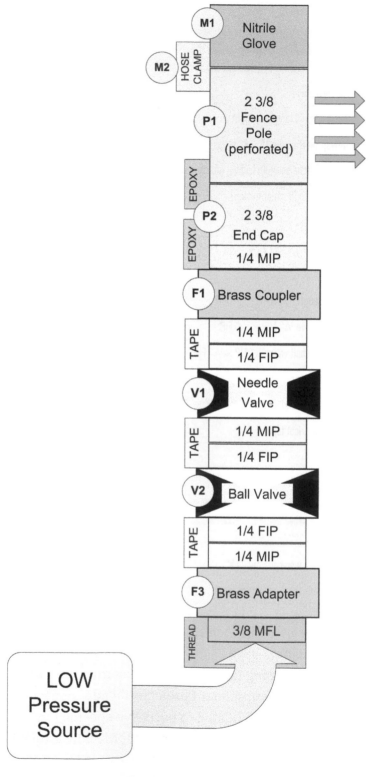

FIGURE 6-4: Rubens tube block diagram

Construction

1. Mark and Punch Holes

Run a length of masking tape down a seam on the tube (P1). Run another length of masking tape down a nearby seam so that it overlaps the first strip of tape. Better yet, use a narrower roll of tape for the second strip and use the edge of the first strip of tape for alignment. Use a different color of tape to make the next steps easier. (See Figure 6-5.)

The edge of the top strip of tape is our registration line.

Starting at 8″, mark lines at ½″ intervals parallel to the registration line. We'll end up with 100 marks that we will make into jets. (See Figure 6-6.)

FIGURE 6-6: Mark the locations for the jets.

At each of the marks on the registration line, use the center punch to make a tiny divot for the drill bit. (See Figure 6-7.)

Drill a $1/16″$ hole in each marked location. (See Figure 6-8 on the next page.) A smaller hole would actually perform better, but $1/16″$ is the smallest size that most common drills can hold in their chuck. We'll try some experiments later to reduce the hole size.

If the drill travels a little, stop before you let it eat much tape. Smooth the tape down and use the center punch to reestablish the marking divot. The bit will travel if you don't start in the divot or the tip is loaded up with chips. Take your time and clean your bit often to get good results.

FIGURE 6-5: Use more tape to create a lateral line.

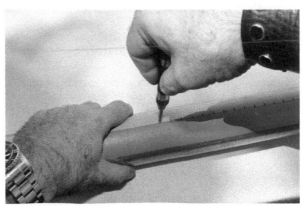

FIGURE 6-7: Make a divot at each mark.

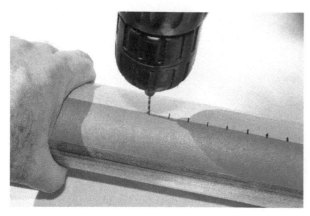
FIGURE 6-8: Drill a hole for the jet at each mark.

It's also realistic to expect to break one or more bits. I broke three bits drilling the 100 holes for the tube. Hardware stores sell these bits in packs of two or three for this very reason.

2. Build the Stands

The stands for the Rubens tube are very simple. We're going to attach two sections of 2×4 underneath the tube. You can use any length of board you like, but a 12″ section for each leg is more than enough. We're going to use a part designed for chain link fencing called an *adjustable wood adapter* to mount the tube to these boards. These adapters have two crown bolts that come up from underneath and keep the adapter from sitting flush on the board. We'll drill two holes where they'll be positioned so that we can mount the adapter cleanly to the board. (See Figure 6-9.)

Use the widest of the step drills or a ½″ drill bit to drill a hole that doesn't go all the way through the board (also known as a *blind hole*.) Angle the drill and rotate the bit around the hole to widen the edges. It's okay for this to look rough, since it will be covered up by the adapter. (See Figure 6-10.)

FIGURE 6-10: Blind holes in the stand board

Though it's not required, I recommend that you paint the boards with a couple coats of spray paint. When the paint is dry, center the adjustable wood adapter on the side of the board so that the protruding crown bolts on the bottom rest in the divots we carved out for them. (See Figure 6-11.)

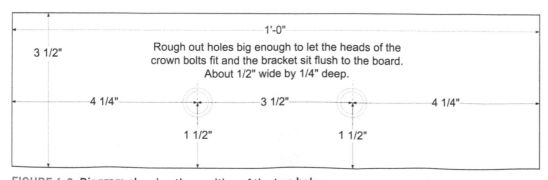

1'-0"

3 1/2"

Rough out holes big enough to let the heads of the crown bolts fit and the bracket sit flush to the board. About 1/2" wide by 1/4" deep.

4 1/4"   3 1/2"   4 1/4"

1 1/2"   1 1/2"

FIGURE 6-9: Diagram showing the position of the two holes

FIGURE 6-11: **Adapter mounted on the board**

On the other side of the board, we will add adjustable-height feet (S3) to be able to level the tube. Read the instructions to get the specific values you'll need for the feet you purchase. For the feet I used, I drilled two $^{23}/_{64}''$ holes 1½" from each end on the centerline of the board. These were blind holes that were 1⅛" deep. To drill a blind hole like this, measure the desired length from the end of your drill bit and wrap a piece of tape around it (so that the depth we want is equal to the amount of exposed drill). This will be your depth guide. As the hole gets deeper, you should withdraw the drill a few times to give the hole a chance to clear the chips out. Continue to drill until the edge of the tape is at the face of the board.

Push the feet into the holes until the nylon bushing is fully inserted. You may need to tap the base of the feet lightly with a hammer to get them to fully insert. (See Figure 6-12.)

You'll construct two of these. When complete, slide one 6" (150 mm) stand onto each end of the tube and tighten the bolts to hold it in place. Be sure the jets are pointed straight up before you tighten.

3. Prepare the End Cap

Use the center punch to mark a location as close as possible to center on the inside of

FIGURE 6-12: **The mounted feet in the board**

the cap. This should be pretty easy to determine by eye. The end cap is a difficult item to clamp without a larger vise than we've been using. If you do have a larger vise, use it to clamp the end cap while drilling. It is possible to hold the end cap safely by hand, but *only if you drill very slowly.* I repeat, only use the drill at a very slow RPM to avoid turning the end cap into a projectile.

Put on your safety glasses and gloves. Hold the end cap with a gloved hand only!

At *very* slow RPMs, drill a pilot hole with a small bit ($^3/_{32}''$–$^5/_{32}''$; the size isn't critical.) Then slowly drill a ½" hole through the center of the end cap (P2). (See Figure 6-13.)

FIGURE 6-13: **Drill the hole for the fitting in the end cap.**

Using Epoxy

Epoxy resin is a powerful adhesive that has the advantage, for our purposes, of not being dissolved or corroded by propane. I must confess that I've had a very hostile relationship all my life with adhesives. For the most part they almost always refuse to do my bidding. Epoxy is the one adhesive I completely trust and feel confident with.

Epoxy comes in a variety of formulations, but we'll be using two-part permanent epoxy. This means that the product comes in two tubes, a resin and a hardener, that need to be mixed just prior to use. There are dozens, if not hundreds, of choices when purchasing. Any basic epoxy will work.

To use epoxy, follow the directions on the package. For the most part, that will consist of mixing an equal dollop from each tube and applying to the target. I used the scientific term *dollop* since the amount to use varies depending on how much epoxy you want to end up with. Unless you have a big job, you aren't likely to want to use the whole tube. I tend to use common coins as a reference for how much of each part to squeeze out. Some small projects use a dime-sized dollop of each part, some use a quarter-sized dollop.

Here are a few tips for using epoxy;

- You won't regret it if you put on a pair of latex or nitrile gloves any time you handle epoxy. Epoxy resin is pretty safe (depending on the kind) but many types of hardeners may present some risks. Gloves are always a good idea.

- Rough up the surface to be epoxied if it's really smooth. A few strokes with a file or heavy-grit sandpaper will help the epoxy form a strong bond on metal.

- Mix the two parts on a piece of wax paper, tin foil, or some other disposable surface.

- Only mix up as much as you'll use in less than a quarter of the drying time (if the drying time is five minutes, only mix as much as you'll use in two and a half minutes).

- Mix the epoxy completely for at least a full minute. You don't have to beat it like an egg, but you want both parts to be completely mixed.

- Toothpicks are a great disposable tool for mixing and applying small amounts of epoxy (like we'll use in this project).

- Be sure to always put the caps back on the same tubes they came from. If you inadvertently swap caps, you're likely to permanently seal the tubes.

- Let the epoxy cure undisturbed for the full cure strength time specified on the package, usually 24 hours. The package may tell you the joint will support loads in minutes, but resist the temptation.

Before it's mixed, epoxy has all the usual warnings that many household chemicals have. Always keep epoxy out of your eyes. If you get some in your eyes, flush with clean water for 15 minutes and contact a physician. Don't swallow epoxy. If you, someone else, or a pet does so, contact a physician (or veterinarian) and keep hold of the package so you can show it to him or her. Keep epoxies out of reach of children.

You may need to use a step bit as a reamer to get the hole circular. Do this by hand, not with the drill! (See Figure 6-14.)

From the outside, test-fit the ¼" MIP × ¼" MIP fitting (F1) into the end cap so that its threads protrude into the bowl on the inside of the cap. Remove the fitting and give the area directly around the hole on the cap a few strokes with some heavy-grit sandpaper.

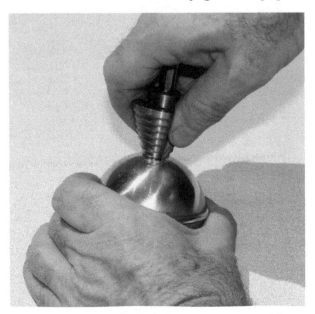

FIGURE 6-14: Ream the hole by hand with a step bit.

Before you go any further, clean the fitting and the cap with some rubbing alcohol or a degreaser. We want the materials to be clean and free of residue. Pour some rubbing alcohol onto a paper towel and run it back and forth on the threads of the fitting. You don't have to go overboard trying to clean it (I can get obsessive about such things); just remove any obvious materials or grease that will compromise the epoxy. When you're done, place the fitting back into the end cap.

Mix a batch of epoxy using a dollop about the size of a dime (~20 mm) for each part. Apply the epoxy around the outside of the end cap where the fitting goes through. Be sure to leave the center of the fitting clean where a wrench will fit. Later we will want to be able to tighten against this fitting without twisting it in the end cap later. (See Figure 6-15 on the next page.)

After giving this a couple hours to firm up, mix up the same size batch of epoxy and seal around the fitting on the inside of the cap. Be careful not to disturb the epoxy on the outside. Set the cap in a cup or something similar (fitting pointing down) that supports it without disturbing the fitting. Let this dry and harden for 24 hours undisturbed.

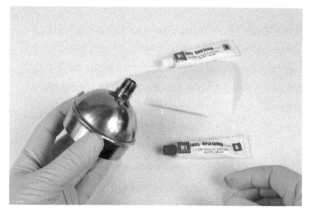

FIGURE 6-15: Epoxy the fitting into the end cap, leaving the clamp-able section of the fitting clean.

When the epoxy has fully cured, tape the outside threads with yellow Teflon tape. Take tight hold of the fitting with a crescent wrench. We are going to attach the rest of the fitting stack and we don't want to transfer torque to the epoxy. Thread the needle valve (V1) onto the taped threads. Tighten using another crescent wrench.

Tape the threads on the needle valve with yellow Teflon tape. Hold the needle valve with a wrench and tighten the ¼″ FIP × ¼″ FIP ball valve (V2) onto it with your other wrench. Make sure that the ball valve is positioned so that the handle points away from the needle valve when open (otherwise they'll conflict). You'll probably also want to make sure that the handles of both valves are aligned. Now tape the threads on the ¼″ MIP × ⅜″ MFL adapter (F3) and tighten it into

the ball valve holding a wrench on each fitting. (See Figure 6-16.)

4. Build the Speaker Connection

Speakers move back and forth as they create sound, compressing and rarefying air in successive waves. The design problem we want to solve is to transfer as much of this compression and rarefaction to the tube as we can. We need to couple the speaker to the end of the tube, isolating it from the propane. Rubber isn't resistant to propane, so ideally we'd use a nitrile membrane, which we have

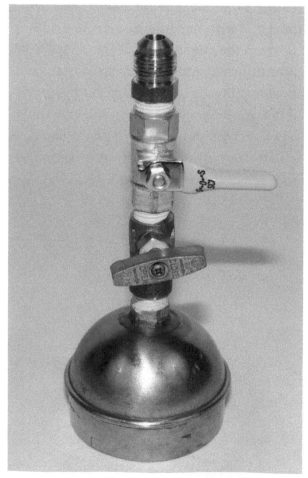

FIGURE 6-16: The fitting stack on the end cap

in the form of nitrile gloves (M1). It's possible to buy sheets of nitrile, but there's really no need. Nitrile gloves are common, cheap, and useful for lots of other things. A box of a hundred 5-mil gloves costs less than a sheet of nitrile and the gloves are great to have around the house, shop, or studio.

In my attempt to both restrict the cost of this project and allow experimentation with different speakers, I didn't want to require a specific speaker. Instead, I'm recommending any powered computer speaker system that includes a subwoofer; even better would be one with a powered subwoofer (A1). These come in a variety of shapes and sizes, providing a number of different configurations of speaker size, vent hole layout, box structure, and so on. The Rubens tube really only responds well to relatively low frequencies (this design shows off its best waves between 420–480 Hz) so the subwoofer provides two useful features. The first is that it does a better job of replicating the low frequencies than the regular speakers, and second, it is typically a larger speaker and moves a greater amount of air.

Subwoofers usually have a hole somewhere in their enclosure. This allows the air behind the speaker to move without becoming compressed in the box. This air is moving 180° out of phase with the air in front of the speaker, meaning that when the air in front is compressed, the air exiting the port is rarefied, and vice versa. For our purposes, it doesn't really matter whether you use the speaker or the port. You may want to experiment with trying each of them. Just be sure

not to use both at the same time, since they would cancel each other out.

I'm going to describe an extremely low-cost method of fabricating a connection that will allow you to make as many custom connectors as you want. We will slide one side of this connector over the nitrile-covered end of the tube and push the other end of the connector flush against the speaker or vent hole (not both) on the subwoofer.

There are many ways to fabricate such a connection. We could build a box with the speaker on one side and the tube entering from another. We could 3D-print a custom fitting. We could use sheet metal fabrication techniques. We could even buy a plumbing reducer and try to mate it with our tube and speaker. All of these are viable approaches and I recommend experimenting with as many as you can. But I want to offer an extremely old-school technique—literally thousands of years old. In many ways it's the original plastic. I'm talking about papier-mâché.

Papier-mâché allows us to build a strong lightweight form using cheap and easily acquired materials. There are many recipes for papier-mâché, and you're welcome to use any of them if you have a favorite. I'm going to offer my favorite: paper strips with thinned Elmer's glue (or any other polyvinyl acetate–type glue like wood glue, school glue, white glue, or carpenter's glue).

Papier-mâché works best when built up from a frame. While there are many solutions that will work, I've recently been using a material called plastic canvas and happily recommend it for this project. Plastic canvas is an incredibly useful material. It's a plastic mesh and comes in rigid or flexible sheets. (See Figure 6-17.) We need the flexible type for this project. You could also use chicken wire, hardware cloth, or any other material you can bend into a ring and drape wet paper strips over.

FIGURE 6-17: Plastic canvas

We need to make two rings, one the diameter of the 2⅜" (60 mm) Rubens tube and the other slightly larger than the diameter of the speaker. In my case, that was 3 ½" (about 90 mm).

Cut 2" (50 mm) strips of the plastic mesh. In my case, I'd need about 8 ¼" (210 mm) with margin for the tube and at least 12 ½" (320 mm) with margin for the speaker. I wrapped the first one around the tube, firm but not tight, and marked where the overlap was. I wanted the mesh to overlap by at least two squares so that I could tie it together with twist ties (string or wire would have worked, but I always seem to come home from the

grocery store with a pocket full of twist ties, so I like to use those). Tie at a couple of points along the seam to form a ring that will slide on and off the tube, then cut off the excess mesh. (See Figure 6-18.)

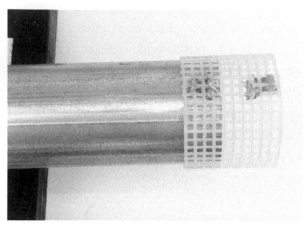

FIGURE 6-18: Checking the fit on the tube end

Wrap the other strip of plastic canvas around the outside of the rim of the speaker. Mark the overlap, tie, and trim.

Now we need to fabricate a reduction from one ring to the other. I used more twist ties to accomplish this. A stiffer wire might have been better, but the twist ties gave me enough structure to support the papier-mâché. Pick eight locations spaced evenly around the rim of the big ring. At each, insert a twist tie through the top hole of the mesh and down as far as the third hole. Bend it back up and twist it around itself to keep it in place. This assists you in getting a consistent length for each tie. Now do the same thing at the other end of each tie onto the small ring. This will be the frame that we'll papier-mâché onto. (See Figure 6-19.)

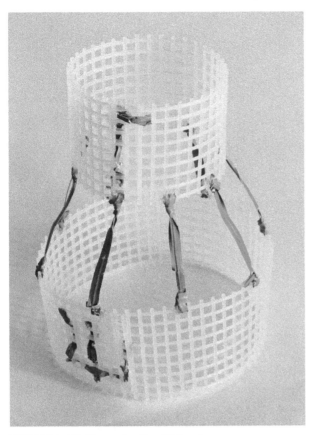

FIGURE 6-19: **The reduction fitting frame**

Papier-mâché is one of the few things I *don't* recommend wearing gloves for. It's messy fun that has a lot of tactile aspects. So lay down a couple sheets of newsprint to protect your surface and get ready to get your hands goopy.

Dip each strip into the glue mixture and squeeze it out between your fingers. You want each strip to get fully coated, but not to have globs of liquid glue on it. Start with strips around the end of each ring closest to the reducing ties. This will help anchor the strips that will cover the conic part.

Lay strips onto the conic section so that they curve around and down toward the larger ring. They should overlap the ring just a little. (See Figure 6-20.)

Feel free to add additional ties; eight is the bare minimum that will work well.

Prepare a bowl of glue mixed with water. If you're using a thick white glue, this is a 50/50 mix. Tear up some newsprint to get 1" (25 mm) strips that are about 4" (100 mm) long. You'll choose how many you'll need, but you'll want to put at least four to five layers so prepare eight to nine dozen. We'll also use a bunch of 1" × 2" (25 mm × 50 mm) strips, so tear up a couple dozen of those, as well. Make sure they're all pre-separated; once we start, you won't be able to easily pull them apart if they're stuck together.

FIGURE 6-20: **Covering the conic section**

Continue adding strips to the conic section until it is completely covered. You may need to add short strips to cover open areas between the conic section and the small ring. If you can't get the strip to lie flat, try a shorter strip. Use long strips wrapped around the two rings to cover and anchor all the ragged edges.

Cover the frame completely and then do it all again. Papier-mâché relies on many layers, so we'll want to do at least four complete coverings of the entire frame—six or seven layers would be great. We'll also papier-mâché the inside of the frame. Don't put too many layers on the inside of the small ring where it will slide onto the tube; we don't want it to fit so tight that we can't get it over the membrane.

It will help to occasionally hold the connector up to a light and look through the inside to see if there are any thin spots that need additional strips. It's also useful to squeeze the inside and outside between your fingers to make sure there aren't glue pockets.

Don't worry about placing the strips in an aesthetically pleasing manner; you can always sand and paint it when you're done. The primary goal is even coverage and good strips across areas that you want to be strong (like the boundary between the conic section and the rings).

I almost didn't want to paint mine. I'd used torn-up flyers from my local discount import tool store and the connector was covered with a riot of tools, generators, wheels, and other items near and dear to my heart. (See Figure 6-21.)

Let the connector dry for 20–30 minutes and decide if you want to add additional

FIGURE 6-21: The unpainted connector

layers. The connector will be thinnest in the conic section, so additional layers will help its strength. Repeat this as often as desired.

Since I hate to waste things, after the connector dried overnight, I mixed up the remaining epoxy from the tubes I'd used earlier and applied it to the inside of the conic section and the boundary area where it meets the rings. Mixing it in a small paper cup made this easy. Letting this dry overnight added a lot of strength and firmness to that section. This is difficult to apply, except by hand, so absolutely wear gloves if you choose to do this. Be careful: Epoxy mixed in quantity is exothermic, meaning it gives off heat. The amount from two small tubes doesn't get too bad, but you can get badly burned or even start a fire if you're not careful mixing large amounts. Always have excellent ventilation when working with epoxies.

Once the tube has completely cured, you can lightly sand and paint it with an acrylic paint. The last step is to add a foam ring around the end of the tube to help mate it with the speaker. Use adhesive foam sold for insulating windows or doors (M3). Peel off a layer and wrap it around the speaker end of the connector, leaving just a small amount hanging over the rim. The adhesive will hold it in place, but a wrap of electrical tape (M4) will help ensure it won't come off. (See Figure 6-22.)

FIGURE 6-22: Attach foam to the connector.

Safety Note

This is the part where I will tell you three things about Rubens tubes that you almost never read on the Internet sites. As always, I rely on the basic safety principles in the initial chapters to support my statements. Here is the conclusion that I am forced to come to:

DO NOT OPERATE A PROPANE RUBENS TUBE IN AN ENCLOSED SPACE WITHOUT EXCELLENT VENTILATION.

The first safety concern is that it is almost impossible, in the course of operating a Rubens tube, to avoid tramp gas. *Tramp gas* is a term that refers to unignited gas that has escaped your device. Rubens tubes frequently lose flame and release tramp gas from multiple holes or go out completely due to high-volume spikes. They are also releasing tramp gas before and after ignition. While propane is completely nontoxic, there is a fire and, in the worst case, asphyxiation hazard when propane is released into an enclosed space. You'll notice the odorants used in propane when you're around a Rubens tube. This is a clear sign of tramp gas.

Additionally, when you turn off the propane supply to the Rubens; tube, it will go out, but the propane remaining in the tube will slowly mix with air (remember, it's heavier than air, so it won't escape out the top holes on its own). There will be a period when the propane remaining in the tube will be within its explosive limits. Ideally, once it has cooled sufficiently, remove the nitrile membrane and vent the tube completely. Never allow smoking or open flames near your Rubens tube while operating or until it has had a chance to completely vent after it's extinguished.

The second safety concern is that the Rubens tube burns rich. A rich burn is when the ratio of propane to air is below the complete (stoichiometric) burn point (4.2% propane, 95.8% air). Since the tube is filled up with propane (ideally forcing out the air in the process) there isn't much additional air mixed into the propane prior to it being burned while exiting the Rubens tube. Look at various Rubens tube videos online. There are generally yellow flames exiting the holes. This means that water, carbon and, most importantly from a safety standpoint, carbon monoxide result from a Rubens tube burn.

Carbon monoxide (CO) is toxic to humans in concentrations greater than 35 ppm. In a well-ventilated space, a Rubens tube will never come remotely close to this. A car produces wildly more CO than a Rubens tube will, and we walk around them safely all the time. But I still have to caution you to never have an incomplete propane burn in an enclosed space.

The last safety concern with Rubens tubes is that they are a very hot, large, metal pipe. This may sound obvious, but imagine the environment where you will operate a Rubens tube. Will there be people who are unfamiliar with propane and high heat? Will they have the ability to touch the tube? Never underestimate the ability of people to do things that you think are stupid.

If you light up a Rubens tube in front of a crowd of people, there will be someone in that crowd who wants to touch it. They *might* be smart enough not to do so while it's lit, but the tube will stay burning hot for some time after it's extinguished. Always keep people away from the tube until it has cooled completely. Always have heavy gloves on hand for handling the hot tube, if necessary.

Assembly and Testing

Because we're putting together a Rubens tube platform that is intended to allow experimentation with a variety of configurations, these assembly instructions are really just a starting point to get a working device. You may choose to put things together in a more permanent manner once you've experimented with different configurations.

Attach the Propane Source

You could use epoxy for this step and make a permanent connection, but for now try this method of attaching the end cap. Using electrical tape made of vinyl (which is chemically resistant to propane), *not rubber* (which isn't), start about 1″ (25 mm) from the end and take four turns around the end of the tube in a spiral, ending with a complete turn around the end. (See Figure 6-23.)

FIGURE 6-23: **Wrap the tube with tape.**

Slide the end cap onto the end of the tube over the tape and all the way onto the tube. Tightly wrap tape two to three times around the end of the end cap so that the tape is half on the end cap and half on the tube. (See Figure 6-24.)

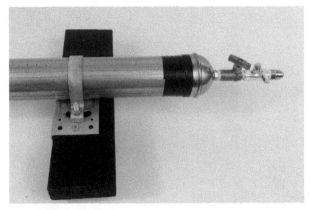

FIGURE 6-24: **The end cap mounted on the tube**

Attach and tighten the low-pressure source to the flare fitting on the end of the end cap stack. Make sure that the needle valve and pressure source ball valve are closed.

Attach the Sound Source

Slide a 5 mil (or thicker) nitrile glove (M1) onto the opposite end of the tube and position it so that a flat area is covering the tube mouth. You don't want the glove to be drum tight. If the glove is too tight, it will serve to resist the transmission of motion from the speaker. Slide a hose clamp (M2) onto the tube about an inch (25 mm) from the end of the tube and tighten. (See Figure 6-25, next page.)

Slide the reducing connector over the end of the tube. Position the subwoofer so that the large end of the fitting completely surrounds the speaker and the foam is pressed up against the cabinet. (See Figure 6-26.)

It's entirely likely that you have to adjust the height of the tube to mate well with the speaker. Turn the adjustable feet to raise or lower the tube.

FIGURE 6-25: The nitrile glove mounted on the tube

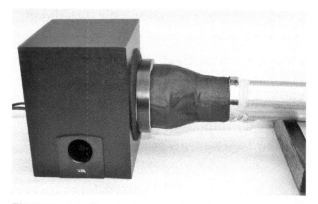

FIGURE 6-26: The speaker positioned for use

Leak-Test up to the End Cap

With the needle valve completely closed, open the cylinder valve all the way and back it off a quarter turn. Open the pressure-source ball valve and use your leak-test kit to test the fittings between the hose and the end cap on the tube. If any leaks are found, shut everything down, tighten, and test again. It's always a good idea to retest the pressure source as well, so go ahead and test it just to be safe.

Operation

Connect an audio source to the input on the speaker. The very best audio to use is from a signal generator or something that will produce pure tones (sine waves). There are free iOS and Android apps for this, so a smartphone or tablet makes an ideal audio source for the Rubens tube. If the subwoofer is powered and has its own volume control, turn it up all the way. Turn the audio source volume (or the amplitude on the wave) about a third of the way up.

There's no point in trying the Rubens tube in anything more than a light breeze. Even a mild wind gust could blow out the jets, so wait for a calm opportunity. Also be sure that the area you intend to operate in has no flammable materials above the tube or that are able to blow onto or over the tube. Make sure that spectators are clear and can't bump or grab the (soon to be) hot tube.

Start with the pressure-source ball valve, the tube ball valve, and the needle valve closed. Open the cylinder valve all the way and back it off a quarter turn. Open the needle valve all the way. Open the pressure-source ball valve. Ignite your fireplace lighter and hold it up to the jets closest to the propane source. Open the tube ball valve. It will take a few moments for the gas to build sufficiently to ignite the jet. If the fireplace lighter goes out, reignite and wait. You may want to wave the lighter closely over the row of jets to assist in getting them lit, especially if there is a small breeze.

Once one or more of the jets ignite, the rest should ignite by themselves. If they don't, use the lighter to get all the jets burning. Let them all burn for a moment; the pressure is likely to continue to build for a minute or two. Adjust the needle valve so that the jets are all burning with a flame no more than an inch high. (See Figure 6-27.)

FIGURE 6-27: The jets ready and lit

Safety Note

If, after 90 seconds, the jets do not ignite, extinguish the lighter and close the needle valve, cylinder valve, and both ball valves. Be *sure* that there are no open flames nearby! Remove the speaker reducer, loosen the hose clamp, and remove the nitrile glove from the tube to evacuate the propane.

Check the following things:

- Verify that the cylinder has propane in it.

- Verify that the glove doesn't have any holes in it (and the hose clamp is tight when you reinstall).

- Verify that the fittings at both ends of the end cap stack have nothing visible blocking them.

- Test the low-pressure source with one of the other projects in the book.

- If none of the other tests find anything, disassemble the stack and check each fitting. It's possible that the needle valve may be faulty. Replace the needle valve if necessary.

If, at any time, there is a risk that something inappropriate will be ignited by the Rubens tube flames, or you need to shut it off quickly for any reason, close the quarter-turn ball valve at the source.

The Rubens tube works better with a smaller resting flame. The 1″ (25 mm) flame (or smaller) will get 3–4″ (75–100 mm) tall when in use. Given our goal of signal visualization, bigger isn't better.

Turn on your signal source and output a 460 Hz (or similar) tone. If you're using a source that has a musical keyboard, try a middle B (B$_4$). The flames should look something like those shown in Figure 6-28.

Getting a good wave on the tube depends on the following factors:

1. Finding a signal that sets up a standing wave in the tube. Adjust with the frequency of the signal.

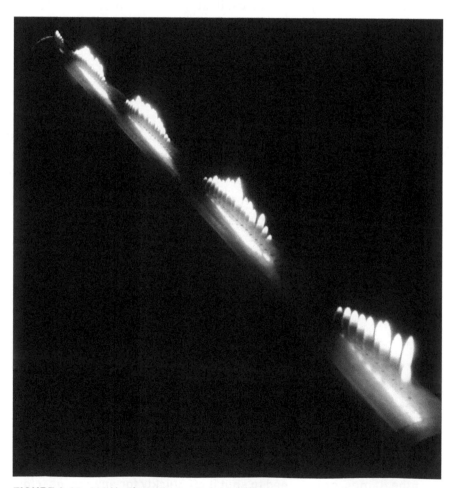

FIGURE 6-28: 460 Hz signal in the Rubens tube

2. The pressure of the propane source (less is usually better). Adjust with the needle valve.

3. The volume of the audio source (mid to low is usually better). Adjust with the volume controls.

4. The decibels (or pressure) transmitted by the audio source to the tube.

The fourth is similar to the third but slightly different. The pressure transferred may have more to do with the seal the reducing connector is making against the speaker cabinet. If air is leaking around the edge, less pressure (fewer decibels) will be transferred. Check to make sure the foam on the connector is up against the cabinet and the speaker is fully enclosed.

Hopefully you're getting a great visualization! If not, slowly shift the frequencies of the audio signal until you start to get a wave form and then adjust the volume and propane pressure to optimize it.

Be conscious of how long you are running the tube to be sure that parts don't get too hot. Only the plastic canvas in the connector is really likely to suffer, and it's fairly insulated. But keep track of all the parts. Have gloves on hand at all times in case you need to deal with the hot tube.

When you finish with the tube, do the following to safely shut it down:

1. Close the cylinder valve.

2. Close the pressure-source ball valve.

3. Close the tube ball valve.

4. Close the needle valve.

5. Put on heat-resistant gloves.

6. Verify there are no open flames nearby.

7. Unwrap the tape from the end cap and remove it to vent the tube.

8. Be sure that no one tries to touch the tube until it has cooled down.

Experiments

END CAP REFLECTOR   Since the signal must bounce off the end of the tube to set up a standing wave, how it reflects will have an impact on the performance. The end cap we're using is a dome. This acts as a lens and is not the best reflector surface. I found that the larger metal discs sold at hardware stores as roofing nail caps fit exactly on the end of my tube. I perforated one with some holes (larger than the jet holes) to let the propane through. (See Figure 6-29.)

I held the reflector against the end of the tube as I wrapped the vinyl tape slightly over it. The edge of the tape held it in place as I pushed the end cap onto the tube end. This served as a flat reflector for the signal and I saw much smoother performance from the tube. If you can't find roofing caps that fit, you can sometimes find a flat end cap. (See Figure 6-30.)

FIGURE 6-29: Perforated end cap reflector

You'd have to get an additional fitting and drill this out, but it's an interesting alternative. You could also try heavy tin foil stretched across the end of the tube, perforated and held in place by the end cap (be careful to seal this well with tape and avoid leaks).

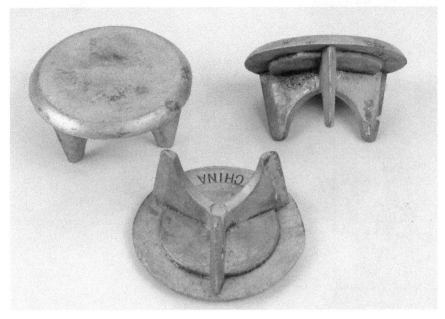

FIGURE 6-30: Flat end cap

ADD INSULATION BEHIND THE REFLECTOR. Try adding some fiberglass insulation behind the reflector to deaden that area acoustically. Fiberglass insulation is nonflammable, so it's safe. Don't use so much that you plug up the propane inlet. Use gloves when handling the insulation to avoid skin irritation.

TRY SMALLER HOLES. Different references on Rubens tubes suggest that smaller holes can perform better. I've had mixed results with this, but you can modify the size of your tube's holes using aluminum foil tape and a needle. Run a strip of foil tape over the existing holes and gently pierce the holes with a pin or needle (try different size needles and compare).

USE THE SPEAKER PORT INSTEAD OF THE SPEAKER. As noted earlier, most subwoofers have a port that puts out air that's 180° out of phase with the speaker. Fabricate an expansion connector to try using this source.

USE A MANUAL AUDIO SOURCE. Rubens used bells as his source. Try using a bell, tuning fork, or musical instrument to create a standing wave. While it's extremely difficult to do, some people can even sing a pure tone!

The High-Pressure Source

7

SO FAR, THE PROJECTS we've been building have relied on a low-pressure regulator. This typically means that we've been receiving propane at about ½ psi. While lots of interesting things are possible with propane at that pressure, even more interesting things are possible with lots more.

High-pressure regulators are available in a variety of ranges. Most of the ones sold to consumers are adjustable and provide a range of outputs beginning at 0 psi (closed) and going as high as 100 psi (700 kPa). Regulators with higher output pressures are available, but for our purposes, I'm going to recommend a 0–60 psi regulator (0–500 kPa would be a common equivalent). The top end of that range will provide a tremendous amount of propane and exceed the needs of the projects in this book. If, after using a 0–60 psi regulator, you want to go purchase a more powerful one, I assure you that you'll be hooked enough on propane work that you'll be happy to have more than one regulator. (See Figure 7-1 on the next page.)

Pay close attention when shopping for regulators. Many regulators sold as "high-pressure adjustable regulators" are only 0–10 psi. Another common range is 0–30 psi. The term *high pressure* just means more than ½ psi, so check the specifications carefully to make sure you're buying the one you really want.

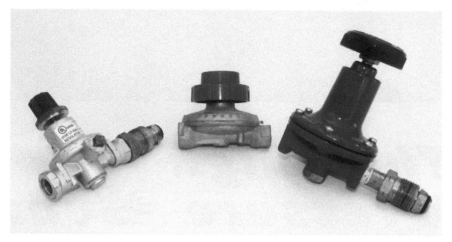

FIGURE 7-1: A variety of high-pressure adjustable regulators

Safety Note

The amount of propane produced by even the smallest of high-pressure regulators can cause injury or death if mishandled. Always follow basic safety practices when working with propane.

- Never draw propane directly from a cylinder without a regulator.

- Always leak-test your system.

- Always include a quarter-turn ball valve close to the regulator as an emergency cut-off source.

- Never allow liquid propane to enter equipment designed for propane vapor.

- Never allow cylinders designed for vapor to be on their side or upside down.

- Always replace, never repair, broken regulators.

High-pressure propane expanding out of the cylinder becomes extremely cold. Be careful and never let allow any skin to be in the path of a high-pressure jet or venting of propane. Never allow ignition sources (other than the ones under your control) within 15′ (4.5 m) of a flame effect.

You can swap out the low-pressure regulator from the low-pressure source to create the high-pressure source, but you'll have greater project potential if you construct a new source. Since the connection from the ball valve to the regulator is taped, it's much nicer to leave it joined than to repeatedly thread and unthread it.

When working with high-pressure systems, it becomes useful, sometimes essential, to have an idea of what pressure you are working at. We will add a 0–100 psi (0–700 kPa) gauge at the regulator to tell us the pressure the regulator is producing. Most high-pressure regulators have a ¼″ or ⅛″ FIP fitting for a gauge built in. I advise buying one that does. If yours does not, you can add a tee fitting after the regulator and plumb a gauge into the system that way.

Some gauges are sold as "liquid-filled." They are generally filled with oil, silicone, or glycerin. Liquid-filled gauges have a number of advantages; they withstand vibration better, are somewhat easier to read, and don't fog up with condensation. However, liquid-filled gauges generally cost more than dry gauges and some are designed to operate in an upright position (forcing the issue by directing the user to snip the rubber nipple on top). Dry gauges are adequate for most flame effect work and handle load-in and load-out better than a leaky gauge that drips glycerin all over your kit. Purchase a dry or fully sealed liquid gauge for your propane work unless the project is rarely going to move.

The regulator should come with the appropriate fitting for the cylinder (POL or OPD). Some regulators do not; check carefully before purchasing. If the regulator does not have the fitting for the cylinder, you may have to purchase a POL × ¼″ MIP adapter. I recommend POL since all cylinders should accept the older POL fitting. Larger cylinders, such as the 100 lb cylinder, frequently do not have the new OPD style fitting, so yet another adapter would be needed if the regulator had only an OPD fitting. This is less of an issue with low-pressure regulators since they rarely require more propane than a 5 lb cylinder can produce. However, a regulator operating at 60 psi or higher can drain a 5 lb cylinder in a surprisingly short time. Operating booshes and other large flame effects frequently leads to the purchase of at least one 100 lb cylinder.

Keep in mind, after releasing significant amounts of propane from a cylinder, the cylinder pressure can easily fall below the rated output of the regulator. If your 60 psi regulator is only putting out 20 psi, check for ice on the cylinder (and wait for it to warm back up!).

Parts

The schematic for the high-pressure source is very simple; note the addition of the pressure gauge. (See Figure 7-2.)

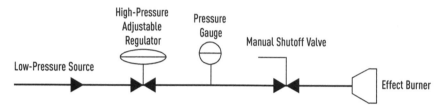

FIGURE 7-2: The high-pressure source schematic

High pressure source parts:

REF	ITEM	QTY
F1	Brass bushing ½" MIP × ¼" FIP*	1
F2	Brass reducing nipple ¼" MIP × ½" MIP	1
F3	Cylinder adapter ¼" MIP × POL	1
G1	Propane cylinder	1
G2	10′ High-pressure propane hose ¼" MIP × ⅜" FFL*	1
M1	0–100 psi gauge ¼" MIP	1
R1	0–60 psi adjustable high-pressure regulator	1
V1	Gas-rated ball valve ½" FIP × ⅜" MFL	1

* Some hoses come with a ⅜" MIP fitting instead of a ¼" fitting. This is fine as long as you change the bushing to match the fitting on the end of the hose. As always, there are many ways to put this together correctly. If your hose had a ⅜" FFL fitting on both ends, you could use a ½" FIP × ⅜" MFL ball valve and skip the bushing altogether. As long as the combination of parts is all gas and pressure rated, it's okay to use a different configuration to achieve the same end.

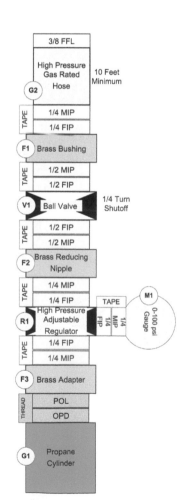

FIGURE 7-3: The high-pressure source block diagram

The block diagram for the high-pressure source is also very simple. (See Figure 7-3.)

Construction

1. Assemble Fittings onto the Regulator

The regulator you purchased may have come with fittings already attached; if so, just skip the parts that are already assembled. I'm going to describe the setup as if the regulator (R1) has no fittings attached. (See Figure 7-4.)

The vast majority of adjustable high-pressure regulators sold in the United States have ¼″ female fittings on all orifices. This is interesting because it says something about the downstream component diameters. With the exception of accumulator systems and long hose runs that we'll describe later, any fitting or pipe that has a diameter larger than the smallest orifice in the system doesn't buy you any additional throughput. Typically, the component with the smallest diameter will be the ¼″ output on the regulator. Spending extra money to purchase large-diameter hose or a ¾″ ball valve doesn't necessarily gain anything. If ¼″ gas-rated components are available for you, they are fine for use in the systems we're building. However, in the United States, ½″ components tend to be the most common and cost-effective purchases. I'm building projects in this book using both ½″ and ¼″ ball valves to demonstrate both approaches.

The first fitting to add is a connection to the cylinder. I describe above why I'm recommending a POL fitting, and based on that, you need a ¼″ MIP × POL adapter (F3). Wrap the ¼″ male end with yellow Teflon tape and tighten it securely into the input of the regulator.

Many regulators come with a ¼″ MIP plug in the gauge fitting. If yours has one, remove it and clean out any hardened pipe dope. Avoid getting scraps of dry dope in the regulator while you do this. Tape the male end

FIGURE 7-4: High-pressure regulator without fittings

of the gauge and tighten it into the gauge fitting on the regulator. Pay attention so that the face of the gauge ends up in a position where you can easily read it.

Lastly, tape the ¼" male end of the ¼" MIP × ½" MIP reducing nipple and tighten it into the regulated output of the regulator.

2. Attach the Ball Valve and Hose

Tape the ½" end of the ¼" MIP × ½" MIP reducing nipple and tighten on the ½" gas-rated ball valve. Be sure to get the handle of the valve placed on the upper side so that it is easy to reach and see. (See Figure 7-5.)

FIGURE 7-5: High-pressure regulator with fittings

FIGURE 7-6: The completed high-pressure source

Tape the ½" MIP end of the ½" MIP × ¼" FIP brass bushing and thread it into the ball valve. Tape the ¼" male end of the hose and thread it into the bushing.

That's it! Construction of the high-pressure source is complete. (See Figure 7-6.)

Testing

Testing is easy, especially since without hooking the source up to something, you are limited to testing the regulator and connection to the ball valve. Be sure the ball valve is closed and the regulator is set to 0 psi. Open the cylinder valve all the way and back it off a quarter turn.

Lightly spray the connection between the regulator and cylinder with your leak-test kit. It's not necessary or desirable to soak the regulator with leak-testing fluid. The regulator shouldn't be damaged by it, but it will get slick and sticky. Concentrate on the fitting and a light misting should reveal any leaks. If any are found, close the cylinder valve, tighten the fitting, and repeat the test.

With the ball valve closed, turn up the regulator until the gauge reads 50 psi. You have two choices at this point. You could leave the system in this state and monitor the gauge to determine if there is a leak, but that would leave things in a dangerous state. The better choice is to leak-test the fitting between the regulator and the ball valve. Don't forget to test the fitting where the gauge is installed, as well. If leaks are found, shut all valves, including the regulator, then tighten and repeat the test.

Maker Tip

Pro Tip: The small additional purchase of a ⅜″ MFL plug will greatly extend the testing. Tighten it into the end of the hose and you can leak-test the ball valve–to-hose fittings. The plug is also a great thing to leave in the end of the hose when the source is not in use. I buy plugs or caps for all the open fittings on my projects. It's cheap insurance that you won't regret.

When testing is completed, close the cylinder valve and, with the end of the hose pointed in a safe direction, open the ball valve to expunge the vapor in the system (remove the plug beforehand if you followed my pro tip). Close the regulator all the way and close the ball valve.

8

The Venturi Burner

CONCENTRATED HEAT SOURCES ARE one of humankind's first tools. Makers of yesteryear considered casting, forging, soldering (old-school style), and hammer-welding metal key skills. Adding a high heat source to your toolkit opens up a wide range of opportunities. While charcoal and waste oil are great options for generating heat, propane stands out as a convenient, clean, easily controllable fuel. There are limits: without effort it's difficult to get a propane-fueled forge or foundry above 1500°–1600° F (about 800°–900° C), but that still makes melting aluminum or heating iron to a bright cherry red no problem.

The amateur community has created dozens, if not hundreds, of propane burner designs and variations. High praise is rightly owed to Ron Reil (ronreil.abana.org) whose *Forge and Burner Design* website has been at the center of DIY propane burners for years. Ron has generously shared his designs, encouraging and incorporating improvements and ideas contributed by the community. Time spent on his website will be invaluable if you're building this project. Tuning methods, troubleshooting tips, and burner enhancements are all available there.

The project in this section is a combination of the Reil "EZ-Burner" and a number of the suggested improvements, most notably Robert Bordeax's modification to the jet tube mounting, the Tweco tip suggestion, and an easy axial choke I've added. These combine to make an extraordinarily simple and effective burner. I've used mine for years now, often considering whether to build myself a new or better one, but never needing to.

The Venturi burner mixes air with propane to provide, once lit, a reducing, neutral, or oxidizing flame. Different flames provide different advantages (or disadvantages) for various tasks. Many burners take a forced-air approach using a fan or a blower to create the desired mixture. The Venturi burner doesn't need any electricity or external components to achieve its mix.

The Venturi effect is the reduction in pressure that occurs when the propane is forced through a reduced-diameter pipe. The reduced pressure creates a draw at the mouth of the bell and sucks in air. This mixture also drops in pressure when entering the smaller pipe and draws in even more air.

Changing the pressure of the propane released through the jet changes the amount of air drawn in so that the overall output can be managed by adjusting the output of the regulator. The choke also assists by attenuating the air flow and making the mixture richer or leaner. The combination of the choke and propane pressure allows for a wide range of BTUs and combustion types.

I use this burner as the heat source for my aluminum foundry. Many makers will use one or more of these burners in a forge to do decorative metal work, blacksmithing, or knife making. Friends of mine have used it as the heat source for their raku trash-can kilns. This design is a rugged, versatile tool that is easy to use and adapt to new purposes.

You'll need to read the "Enhancements" section at the end of the chapter to build a flare and use the burner in open air. You can use the burner in a crucible foundry as is and it's an absolute beast! Hearing it roar and watching a crucible full of aluminum turn bright silver and cherry red at the same time is one of life's true joys!

Parts

The schematic for the Venturi burner is extremely basic (see Figure 8-1).

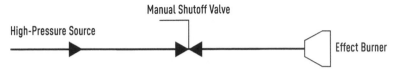

FIGURE 8-1: The Venturi burner schematic

Burner parts:

REF	ITEM	QTY
F1	Brass adapter ¼″ MIP × ⅛″ MFL	1
F2	Brass bushing ¼″ MIP × ⅛″ FIP	1
F3	Brass cap ⅛″ FIP	2
F4	Brass tee ⅛″ FIP × ⅛″ FIP × ⅛″ FIP	1
F5	Brass nipple ⅛″ × 2″	2
F6	Brass nipple ⅛″ × 1 ½″	1
V1	Gas-rated ball valve ¼″ FIP × ¼″ FIP	1
T1	Tweco 14T tip *	1
P2	Black iron nipple 2″ × 3″	1
F7	Black iron reducing bell 2″ FIP × ¾″ FIP	1
P1	Black iron pipe ¾″ × 8″	1
P2	Black iron nipple 2″ × 3″	1
F8	Copper adapter ¾″ FIP × ¾″ cup	1
C1	Hose clamp 2 ½″	1
C2	Metal roofing cap 2 ½″	1
C3	Draw bolt ¼″-20 × 3 ½″	1
C4	Metal hanger strap ¾″ × 20″	1
C5	Nut, washer, and lock washer ¼″	2

* Tweco is a brand name of an excellent MIG welding tip. The 14T has an internal diameter of .044″. Personally, I've substituted much cheaper .045″ MIG tips from a local discount import tool store. Both have ¼-28 male threads and will work for this project.

Shop carefully for the ¼" ball valve. Make sure that the valve you choose is rated on the body and handle, ideally with the older WOG rating, as well as with the newer CSE rating (but at least one of those must be present). If you can't find a ¼" valve, you can use a ½" valve if you change the related fittings.

As the Internet eats away at local plumbing supply stores and erodes the stock carried by big-box hardware stores, parts outside a small range of commodity sizes are increasingly difficult to find. A case in point is the 2" × ¾" reducing bell. A few years ago, I purchased one locally for a burner, but I surveyed all my local suppliers when I started this book and the best I could find was a galvanized 2" × 1¼" reducing bell and a 1¼" × ¾" galvanized bushing. As a result, I had to purchase the desired bell online. You may live in a big enough city that there are still plumbing supply stores where you can wander dusty aisles and find treasures. If so, count yourself lucky and try to help keep them in business. You'll miss them when they're gone!

You can substitute many parts in a design like this, but there are some constraints. Galvanized is an okay substitute for black iron. However, you should stick to the dimensions and lengths of the burner as they factor into the ratios of air/propane mixture. A reducing bell does a better job of handling the mixture flow than a union with a bushing, but you could probably get away with it if you had to.

Apologies to metric makers, but I'm not confident in the exact metric replacements, so I'm specifying in inches.

The block diagram is broken into the parts associated with the propane flow and the superstructure of the burner: the reducing bell and associated pipe (see Figure 8-2).

Water tuning harness parts:

REF	ITEM	QTY
G1	1-gallon yard sprayer (with misting wand)	1
G2	Brass adapter ¼" MIP × 3/8" MFL	1
G3	Brass swivel union 3/8" FFL × 3/8" FFL	1
G4	Tube ½" ID (6–12")	1
G5	Hose clamps ½"	2

Additional Tools:

REF	ITEM	QTY
A1	Drill bit $^{23}/_{64}$"	1
A2	Drill bit $^{13}/_{64}$"	1
A3	Tap $^1/_4$–28	1

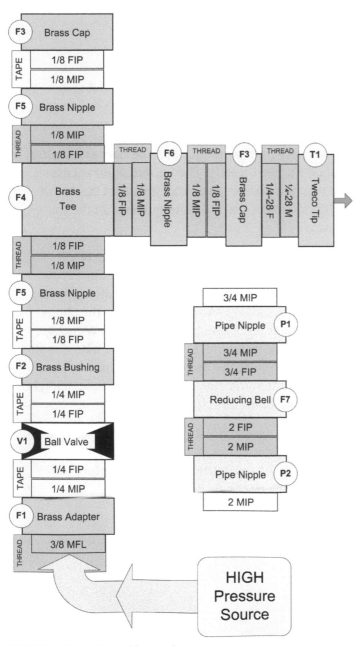

FIGURE 8-2: The Venturi burner block diagram

The larger drill bit will be used to create the holes in the reducing bell for the brass nipples. The smaller drill bit will be used to drill the hole that we'll tap for the ¼–28 MIG tip. We'll discuss tapping in the construction section, but if you don't already have one, you'll need a tap wrench to hold the tap.

Tapping Threads

Cutting threads into the sides of a hole is known as *tapping*. Tapping is a very useful skill. While it can be very frustrating when it goes badly, the basic action of tapping is very simple. Taps are very similar to screws or bolts, but there are a few important differences.

The first difference is that the tap is made of a very hard metal; it has to be harder than the metal you want to tap or it would just grind off its own threads. The second difference is that the tap is tapered. The tip is narrower than the main body so that it can get started in a hole. The third difference is that the tap has channels cut into its length so that the material cut out of the hole can move out of the way.

The rhythm of tapping is different than just threading a bolt or screw into a hole. While the pattern can vary slightly depending on the hardness of the material being tapped, the basic pattern is a quarter turn clockwise and a slight turn back. This pattern allows the tap to break loose the *chips* (material cut from the hole's sides) and have it fall into the channels. When you reverse the tap, you can usually hear a *snick* as the chips break. Without the slight turn back, the tap will eventually get stuck, or worse, break off in the hole. Small taps are notorious for breaking in holes and often require the use of a gentler pattern (an eighth turn clockwise and a slight turn back). Note that the clockwise turn I'm describing is for standard threads; reverse thread taps work the other direction.

Because we're tapping a shallow hole in soft brass, we can get away without lubricant in this project. Tapping normally requires cutting lubricant to work effectively. Cutting oil is sold at most hardware stores for this purpose. Dip the tap in the cutting fluid before tapping, or brush it on while tapping to allow the tap to effectively cut the material.

You can tap a *through hole* (which goes through and out the other side of the material) or a *blind hole* (which goes into, but not out the other side of, the material). Tapping a blind hole is more difficult and often requires removing the tap at intervals to remove the chips. Blind holes often have to be slightly deeper than the intended screw or bolt length so that the tapered end of the hole that the tap cut is below the bottom of the screw or bolt.

Reference charts in books or the Internet can provide you with the appropriate size hole to drill if you're trying to tap for a given bolt or screw size. The hole has to be the size of the screw without threads; this allows the tap to cut threads that will match the outer diameter of the desired screw or bolt, which is the measure by which they are designated.

Taps usually have a square profile on the end to be held and are turned by a tap wrench. Anything that will turn them will work, but the tap needs to be held as closely in line with the hole as possible (usually that's perpendicular to the surface the hole is cut in). Tapping at an angle or oscillating the tap while tapping are the two biggest problems people encounter when trying to tap. Tap wrenches usually have two arms that assist in positioning the tap appropriately (as opposed to holding it with vise grips or a crescent wrench, which would be off to one side). Do your best, and check from multiple angles to keep the tap square to the work.

Construction

This burner, like the rest of the projects in this book, is designed to be constructed with hand tools and a drill. There are aspects of the construction that would be easier with the use of bench tools like drill presses, lathes, and so on. If you own or have access to, and are comfortable with, those types of tools, by all means use them.

The method used in this project is one of two approaches to using a bell and pipe to hold the jet and gas feed. We're inserting the tube through the pipe and the bell tightens down onto it. This requires us to extend the jet tip up into the neck of the bell. The other approach is to drill a hole in the bell for the tube and use the pipe to tighten in place. In this approach, a tapped coupler could replace the tee. Both have advantages, but I chose the pipe method because it allows us to try different jet heights by using different lengths of nipples in the tee.

Some parts of the burner, like the flare, are difficult to build without already having access to a forge and working burner. I'll be offering some "bootstrap" approaches that aren't ideal, but will allow more flexibility from the burner.

1. Tap the Cap

We'll do the hardest step first: tapping the hole in an ⅛″ cap (F3). The relatively thin roof of the cap won't allow very many threads to be tapped, but we only need a couple. The MIG tip we'll use doesn't have to hold pressure; it just has to fit firmly in the fitting. The tip has about a ¼″ or less of threads, so it doesn't need a lot to thread into.

Use a small ruler and mark a center point on the flat of the cap. You can use opposite vertices of the hexagonal cap to draw lines and locate the center. Carefully position the point of the center punch on this mark and firmly press the punch so that it clicks and makes a divot. Without moving the punch, do this two to three more times to really make the divot as deep as possible. (See Figure 8-3.)

Drill a pilot hole with any small drill bit (¹⁄₁₆″ or similar). Then, with the ¹³⁄₆₄″ bit, drill the main hole using the pilot hole as a starter. Do your best to keep the drill perpendicular to the fitting. Technically, a ⁷⁄₃₂″ bit is the designated bit for a hole to tap, but the brass is so soft that a ¹³⁄₆₄″ bit is a better choice. A drill press is an ideal tool for this, but with care, you can do just fine with a cordless (or corded) hand drill.

Insert the ¼–28 tap in the tap wrench, tighten it in place, and position the tip into the hole you just drilled. Keeping the tap perpendicular to the cap, slowly turn it a few turns. The first couple turns will be very easy, but subsequent turns will be more difficult as the tap bites into the metal. Since we're cutting a very shallow tap in relatively soft brass, we can skip cutting oil. (See Figure 8-4.)

FIGURE 8-3: Mark the cap with a center punch.

FIGURE 8-4: Tapping the hole in the cap

Once you get the tap to bite, give it a slight turn backward. Continue this until the tap has inserted about half its length into the cap using the "quarter turn clockwise, small turn back" method (see the "Tapping Threads" sidebar earlier in this chapter). Be very careful to keep the tap perpendicular, and check it from multiple angles every twist. When at least half the tap has made it through the cap, remove the tap by gently turning it counter-clockwise. The threads we have cut, as noted, are shallow and soft, so you don't want to damage them with the tap when removing it. The copper tip we'll be threading into it will be much gentler on the brass of the fitting.

When you're finished, test-fit the tip into the threaded hole. (See Figure 8-5).

For ideas on what to do if the tip isn't aligned well once threaded, see "Trouble-shooting" under the "Tuning and Testing" section later in this chapter. Do not decide the tip isn't workable until you've tested it with the water jet. A tip that looks off-center may be perfect once the other factors are taken into account.

There are other ways we could have accomplished the fabrication of the jet. We could have drilled a jet hole directly into the cap. Don't feel limited by the approach in the book; Ron Reil's website has many approaches and suggestions for other designs.

2. Drill the Holes in the 2″ Section of Pipe

We want the 1/8″ brass tubes to pass through each side of the 2″ × 3″ section of pipe (P2). We'll center the holes for this 5/8″ from the edge on opposite sides of the pipe. This placement means we'll be drilling the holes in the threaded section of the pipe. This will allow the bell, when threaded onto the section of pipe, to tighten up against the tubing and hold the tubing in place. (See Figure 8-6.)

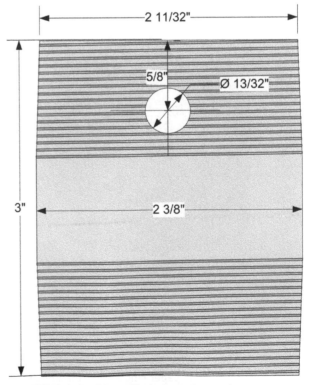

FIGURE 8-6: Position of hole for tube on pipe

FIGURE 8-5: Checking the tip's fit

We want the tube to go completely through the pipe and be as centered as possible. This means that the holes on the pipe should be placed as closely to opposite from one another as possible. In a drill press this is fairly easy; just continue to drill through and drill the second hole from the inside out. With a handheld drill this is somewhat more difficult.

Cut out a 1″ × 8″ piece of paper and wrap it around one end of the 2″ pipe section. Hold it so that it's flush around the edge closer to the center of the tube. The edge around the rim of the tube will *not* be flush, since the threads on the tube are cut with a taper. This is easier if you do it with the tube sitting on a flat surface and wrap around the bottom edge. Tape the paper so that it forms a ring. (See Figure 8-7.)

FIGURE 8-8: Mark a line on one side of the pipe.

Making sure that only the edge closest to the middle of the tube is flush, mark a line on the other side of the threads, 180° away from the first mark. Be careful; if you hold the paper flush against the tapered threads, the edge of the fold will curve around the threads.

You can mark each line at a point ⅝″ from the edge of the pipe to identify the location to drill each hole. It's likely that on one side of the pipe the location is in a groove and on the other side it's centered on a rise in the threads. The groove will be easy to mark with the center punch; the rise will be harder. You may have to flatten it.

Ideally, your vise will hold the pipe for all the punching and drilling. If not, open the jaws as wide as they will go (without falling out) and place the pipe along them so that the

Remove the paper ring (don't undo the tape) and fold it in half. Wrap the paper around the threads again; this time it will only go halfway around. Use a fine-tip permanent marker or a mechanical pencil and mark a line inside the grooves along the edge of the paper. (See Figure 8-8.)

FIGURE 8-7: Tape a 1″ paper ring around the end of the tube.

vise jaws rest along each side of the pipe and keep it from rolling or moving quite so easily.

Take a heavy nail (8–10 penny), and hammer the point so that you blunt the tip. Center the flattener on the rise where you want to drill and use a series of successively stronger taps with the hammer to flatten the thread at the desired location.

Once the center punch has created good divots at the drilling sites, use a small drill bit such as a 7/64″ to drill a pilot hole through each side.

With the pilot holes complete, move on to drilling the 23/64″ hole. If you don't have access to that size drill bit, you can use a 3/8″ bit, but you may need to ream out the hole slightly for the brass tube to pass through. Being creative with the creation of jigs and work-holding is one of the secret joys of fabrication work! (See Figure 8-9.)

3. Assemble the Jet inside the Feed Pipe

With the exception of the external fittings, we have to assemble the jet while the parts are already inserted into the pipe; otherwise we won't be able to pass them through. But we can prepare the external parts first. Tape the end of one of the 1/8″ brass nipples (F5) and thread the brass cap (F3) onto it.

Prepare the other side by taping the 1/4″ MIP × 3/8″ MFL fitting's (F1) threads and tightening it into the 1/4″ ball valve. You'll want to be careful, though, because the ball valve needs to be oriented so that the handle, when open, points toward the flare fitting (otherwise the reducing bell will get in its way).

Tape the 1/4″ MIP × 1/8″ FIP bushing (F2) and tighten it into the other side of the ball valve. Tape one end of the other 1/8″ brass nipple (F5) and thread it into the bushing.

Slide a tube into each of the two holes drilled in the 2″ section of pipe. (See Figure 8-10.)

Thread the tube with the cap on it into one end of the tee. Because any tiny amounts of gas that leak inside will only contribute to the burner's output, we don't need to tape

FIGURE 8-9: Drill the main holes for the tube to pass through.

FIGURE 8-10: Tubes prepped and inside pipe

the threads of the tubes where they mate with the coupler.

Before we thread the other tube into the tee, visualize how the system will work. We want to be able to use the ball valve easily, so we want the handle on top. The tapped hole, once the tip is mated to it, will jet propane vapor down the length of the nozzle tube, so it needs to be aimed down the length of the full burner.

Now thread the tee (F4) onto the brass tube with the ball valve attached to it. Tighten this with wrenches so that the center opening is 90° before the handle of the ball valve, assuming the ball valve is on the right side looking down. (See Figure 8-11.)

If you're left handed and want the ball valve to enter on the other side, just make sure the jet still has the correct orientation when you're done with reversing the directions.

4. Assemble the Burner

Thread a 1½″ nipple (F6) into the tee so that it will aim down the nozzle. Thread the cap onto it and the MIG tip into the cap. Do not overtighten! This should be as tight as you can easily make it by hand. We don't want to strip the threads. If the tip feels loose, give it a few wraps of Teflon tape to help it fit better.

Thread the 8″ section of ¾″ pipe (P1) into the reducing bell (F7). Tighten hand tight; you don't need to tape this joint; any leakage will become additional air sucked into the burner. You do, however, want to make sure it's tight

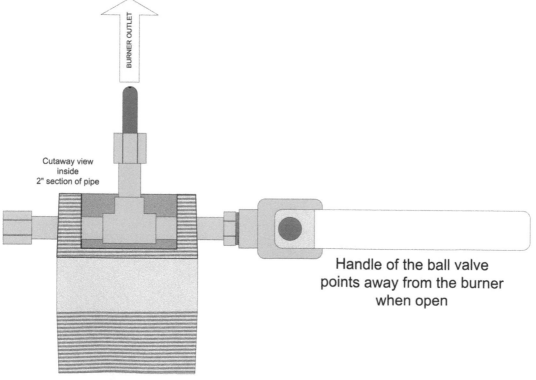

FIGURE 8-11: Jet and ball valve orientation diagram

enough that there is no risk of it coming apart accidentally.

Thread the bell onto the end of the 2″ section of pipe (P2) so that it touches firmly up against the brass tubing. While we don't need to tape this joint for leakage concerns, tape does help stiffen the bell on the threads. Hold the tube in place and tighten the bell as tight as you can. You may want to use a pipe wrench or vise to tighten the bell enough to lock the tube.

Look down the open end of the 2″ pipe to verify that the coupler is centered in the tube. We'll do a special tuning step to get it perfect later, but verify that the jet is pointing down the bell and ¾″ pipe. You should be able to look down the ¾″ pipe and see the opening on the end of the MIG tip. (See Figure 8-12.)

Once you have the tuning completed, if you didn't tape the bell's threads, it's a good idea to put some Loctite or other threadlocking adhesive on them so that the bell doesn't easily slip and allow the jet to move. I also added some putty to the openings where the tube enters the pipe to help keep it firmly in place. Once you get everything tuned, you can use epoxy putty to lock the tube in place.

5. Attach the Choke

Many people construct burners with no choke at all. However, the choke adds an ability to tune the mixture and achieve a better burn under a wider range of propane pressures. You may have seen a choke in a carburetor at some point. This is frequently a disc in the pathway of the air intake that turns in order to cut off air flow or let it pass. This is a solid design, often called a butterfly choke, that many people use in their burners.

To achieve the maximum performance, the air entering the back of the burner should be as laminar as possible until it mixes with the propane. *Laminar*, in this case, means that it flows smoothly with as little turbulence as possible. Once it combines with the propane, turbulence in the nozzle pipe aids in mixing, which is beneficial. Adding a choke that doesn't disturb the incoming air flow is a little bit tricky.

The best approach is an axial choke. This means that you cover the back of the burner completely with the choke and when you lift it, you do so in line along the center channel of the burner. This allows the air to enter and flow evenly around the edge, as if you have

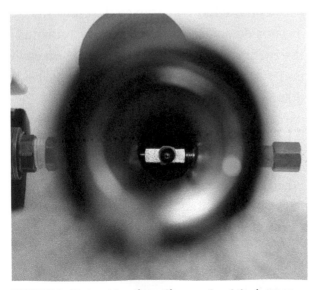

FIGURE 8-12: Looking down the nozzle of the burner

carefully just barely lifted the lid on a pot. Many axial designs require more fabrication than is feasible with hand tools, so I'm going to share the super-simple choke I use on my burners. There are many ways to improve it, but it's worked so well for me that I've never needed to.

We're going to use ¾" hanger strap (C4) to make an arbor to mount the choke. My dad used to call this stuff *plumber's tape*. It's a strip of 28-gauge galvanized metal with holes. It's viciously sharp, so beware! The holes on most hanger straps alternate between large and small holes.

Cut a length of strap just longer than 20" starting at a small hole (cut across the hole). Double it over so that it is about 10" long. We want a large hole to be in the center of the 10" strip, so you will fold it at one of the holes.

Unscrew the threaded bar from the draw bolt (C3). Remove the unthreaded bar (we won't need it). Screw the threaded bar all the way to the top of the draw bolt. Slide the end of the draw bolt through the center hole on the hanger strap.

At this point you can tape or epoxy the tapped bar to the hanger strap so that the draw bolt, when turned, moves up and down through the strap. (See Figure 8-13.)

Drill a ¼" hole in the center of a 2½" metal roofing cap (C2). Thread a ¼" nut, lock washer, and washer (C5) onto the end of the draw bolt; leave about ½" of bolt showing below them. Push the end of the draw bolt through the hole in the roofing cap. Thread a ¼" washer, lock nut, and nut onto

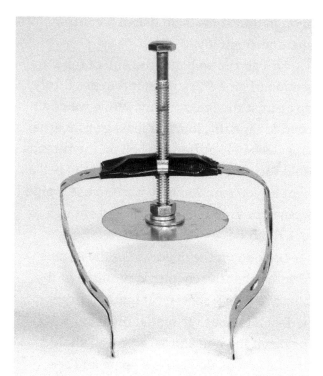

FIGURE 8-13: Draw bar and hanger strap

the end of the draw bolt and tighten so that the roofing cap is held in place on the end of the bolt. (You can see how this looks in Figure 8-13.)

Bend the hanger strap so that it forms am upside down U around, but not touching, the roofing cap. Now bend the last 1" on each end of the strap inward and then the last ¾" down (also shown in Figure 8-13).

Loosely fit the 2½" hose clamp (C1) around the open end of the 2" section of pipe. Slip both ends of the hanger strap under the hose clamp and center them on each side of the pipe. Tighten the hose clamp until the strap is held firmly against the edge of the pipe on each side and the roofing cap is centered over the end of the pipe. (See Figure 8-14.)

FIGURE 8-14: The axial choke

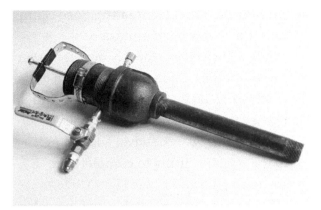

FIGURE 8-15: The completed burner

By rotating the head of the draw bolt, you will cover (or uncover) the opening and attenuate the amount of air entering the burner. When correctly positioned, the roofing cap should completely cover the end of the pipe when it is fully extended.

With this, the burner is complete! (See Figure 8-15.)

Tuning and Testing

The Venturi burner works best when the jet of propane passes directly down the length of the ¾″ pipe and exits dead center. This provides the optimum mixture of air and propane and creates the best draw of air down the length of the burner. There are two axes of control available to us to achieve this placement; the rotation of the brass tube in the 2″ section of pipe and the side-to-side position of the tube and MIG tip.

Rather than attempting this while the burner is lit, or jetting propane, we will use a tip offered on Ron Reil's website and use water in place of propane. This allows the tuning to occur in a completely safe manner.

1. Construct a Water Tuning Harness

This component will only be used with water, so we don't have to worry about the normal concerns about taping or the use of hose clamps. We'll use a garden sprayer as a portable pressurized water source.

You can use almost any sprayer, but since these are typically used for yard and garden chemicals, I recommend buying a new 1-gallon sprayer (G1). These typically cost less than some of the brass fittings we've been using. Slide a hose clamp (G5) loosely onto the end of a section of tube (G4). Unscrew the misting head on the sprayer wand and insert the end of the wand into the tube. Tighten the hose clamp around the wand head and tube.

Slide a hose clamp loosely over the other end of the tubing. Slide the ¼" MIP end of the flare adapter (G2) into the tube, cover with the hose clamp, and tighten.

Thread the ⅜" FFL swivel adapter (G3) onto the ⅜" MFL fitting to complete the water-tuning harness. (See Figure 8-16.)

If a sprayer is unavailable, you can also hook the harness up to a garden hose. Use a female garden hose to ½" ID barb adapter (make sure the adapter comes with a gasket for the hose). Slide a hose clamp (G5) over one end of the tube (G4). Push the barb end of the barbed adapter fully into the tube. Slide the hose barb over the tube where the barb is and tighten. You can hook this up to a garden hose and use the ball valve on the burner to turn the water on and off.

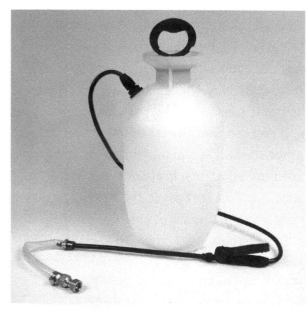

FIGURE 8-16: **The water-tuning harness**

2. Tune the Jet

Tighten the ⅜" FFL swivel fitting on the end of the tuning harness to the ⅜" MFL fitting on the burner. Open the burner ball valve.

There may be water leaks around the harness. If they bother you, tighten fittings until they stop. The leaks will cause no problems for the testing, so if they don't bother you just ignore them.

While holding the brass burner tube roughly in the position that you expect to work, gently tighten the reducing bell to hold the tube firmly in place. Tightening the bell without moving the tube (it will want to twist in place as the bell threads down onto it) is the most difficult part of tuning.

Point the tube in a direction where the water jetting out won't be a problem, pump up the sprayer to pressurize it, and close the trigger on the sprayer wand. Ideally, a stream of water will be shooting directly out of the center of the end of the ¾" pipe. (See Figure 8-17, and note the pretty little mini-flare!)

FIGURE 8-17: **Tuned water jet**

If the water jet is off-center, we can make one or both of two adjustments. (See Figure 8-18.)

The first adjustment will move the brass burner tube side to side in the 2″ section of pipe. The reducing bell should be too tight for this to be easy. You can use *very* light taps with a wrench or tack hammer to move the tube in small increments.

The second adjustment consists of rotating the brass burner tube. You can use a wrench on the hex fittings of the tube, but remember that if the bell is very tight, you don't want to turn a fitting against it in a way that will loosen the fitting on the burner tube. Turn fittings in a manner that will tighten them against the brass tube as well as turning the tube (clockwise).

Once you get the tube in position so that the jet is exiting correctly, mark the tube and the pipe so you have an external visual

indicator of whether the tube is aligned. Move the tube back to its marks if it gets changed. Look down the nozzle end of the burner (you may need a flashlight) and note how the jet looks when aligned. This, along with the registration marks you made, will help you quickly realign the jet if needed.

Troubleshooting

It is not unlikely that the water jet will not be centered when it shoots out of the nozzle. This is due to one or more factors. The hole in the 2″ section may be off-center. The bell, as it traverses the threads, may come down on one side of the tube and push it more than the other side. Alternately, the Tweco tip may be angled in the cap.

Do not despair! If the tube is off-center, shims can help a great deal. My favorite source of shims is a sacrificial set of feeler gauges. These are sold at auto parts stores and are a fantastic source of metal in carefully measured thicknesses. Many are thin enough to cut with scissors, and others need snips. I cut shims from some pieces of sheet metal that I found that were left over from some ductwork. Even shims cut from soda cans would work. Anything firm that will bend around the tube will work as a shim.

Cut the shims just large enough to fit around half the tube and slide them into the hole in the pipe section where you noticed the tube being pushed first. You may have to loosen the bell. It will help to cut them into a wedge shape, sliding the narrow end in under the tube and then tapping the other end in. Or you can use pliers to wiggle it in.

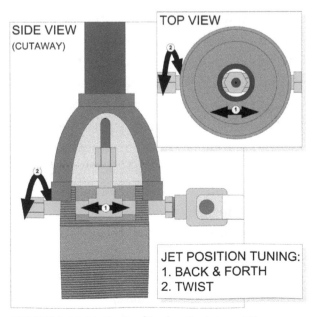

SIDE VIEW (CUTAWAY)

TOP VIEW

JET POSITION TUNING:
1. BACK & FORTH
2. TWIST

FIGURE 8-18: Methods of tuning the jet position

Remember to tighten down the bell when you're done. (See Figure 8-19.)

If the jet is off-center, it may be because the hole was drilled or tapped off-center. It may because the shoulder of the jet where it contacts the fitting may not be even. This can be managed. As noted earlier, you can tightly wrap the threads of the jet multiple times with yellow Teflon tape. Thread it gently into the fitting. Unthread it, do not remove the tape, and wrap four more wraps of tape around it. Thread it back into the cap. We're building up a tighter joint using the Teflon tape.

In the end, you should find that the jet, when fully threaded, has very little motion in it. If there is motion, you can lock it in place. Remove the nozzle pipe so that you're looking at the jet in the neck of the bell. You can use a toothpick or Q-tip (or very small fingers) to place a small ball of underwater epoxy putty (not liquid epoxy) on each side of the jet. J-B Weld makes this under the brand name WaterWeld.

Mold the putty around the base of the jet and repeat the tuning procedure, adding and removing the nozzle pipe to allow adjustments to the jet. Gently squeeze the putty around the base of the jet until the jet is in the correct position. Once you have the jet positioned correctly, leave everything to set and harden in place undisturbed—that will take at least 20 minutes. The downside to this is that it is more difficult to replace jets, but tapping and threading a new cap may be easier than tuning every time you use the burner.

Tuning can be a frustrating task; patience will win the day. If it helps at all, remember that you have water, not roaring flame, shooting out the end of the burner while tuning. Imagine how hard it would be if it was the other way around.

When you're finished, be sure to open the burner ball valve and drain out the water remaining in the brass pipe and jet.

3. Test the Burner

Mount the burner somewhere safe. The flame output must not be pointing at, or close to, anything flammable. The burner itself should not get very hot; any heat should be concentrated near the nozzle. The easiest method for testing will probably be to clamp the burner in a vise with the burner mouth pointing out over an empty space. Give yourself 8′ or more of clearance beyond the burner and at least 3′ on all sides.

Close the burner ball valve and the pressure-source ball valve. Turn the regulator adjustment so that it is closed and be sure the cylinder valve is completely closed. Tighten the FFL (female flare) fitting on the end of the high-pressure source's hose to the

FIGURE 8-19: Shims under the tube

MFL (male flare) fitting on the burner. Attach the high-pressure source to the cylinder if it isn't already.

With the pressure source's ball valve closed, open the cylinder valve all the way and back it off a quarter turn.

Rotate the choke so that the opening is almost completely covered. Open the pressure-source ball valve.

Using your leak-test kit, spray and look for bubbles at the (closed) burner ball valve. Feel free to test the high-pressure source for leaks, as well. If any are found, shut down the system and tighten before testing again.

Get your lighter ready. Do not attempt to light the burner with a cigarette lighter. Use a propane torch or a long-neck fireplace lighter. You are likely to extinguish and reignite the burner multiple times.

With the regulator still closed, open the burner ball valve. Light your ignition source and begin turning up the pressure at the regulator until the burner ignites. Ideally, the burner will immediately ignite and a large loose flame will be produced out the end of the nozzle (and not anywhere else). You can moderate the choke to improve the flame's output.

At lower pressures you are likely to have to rotate the choke to close off more of the entrance for the air. Higher pressures allow you to open up the choke.

Without a flare, the burner has a difficult time operating efficiently in open air. It will work great in an enclosed space like a crucible foundry or a hot forge. To tune the burner for open-air use, see the enhancements section of this chapter and make one of the two

flares. With a flare, the flame should look something like the one in Figure 8-20 when it's ignited and the choke is open.

FIGURE 8-20: **The burner ignited**

The appropriate characteristics of the flame depend, to a large degree, on the intended use. When using it in a foundry without a flare, many people will opt for a slightly reducing (rich) flame to avoid oxidizing their crucible or metal. In most cases, a neutral flame, the hottest flame available from the burner, is desirable. I don't want to try to go very deeply into the vast subject of metallurgy in this DIY book, but I will offer some assistance in tuning the flame. (See Table 8-1.)

TABLE 8-1: **Flame Characteristics**

BURN TYPE		FLAME DESCRIPTION
Rich	Reducing (Carburizing)	Bushy or wispy flame Possibly greenish or yellow flames
Neutral	Stoichio-metric	Light blue, well-shaped inner core Smooth, blue outer flame Roaring sound
Lean	Oxidizing	Bright inner core Long, tapered outer flame Hissing sound

Ron Reil has provided an image of what the intense high-temperature flame with a well-contained envelope should look like. (See Figure 8-21.)

You can shut the burner off by closing the burner ball valve. To completely shut the system down, close the burner ball valve to extinguish the flame. Close the cylinder valves completely, wait 60 seconds (or long enough to let the flare and nozzle cool down so they will not reignite the propane) and then open the burner ball valve to vent the hose. Finally, close the pressure-source ball valve and then the burner ball valve.

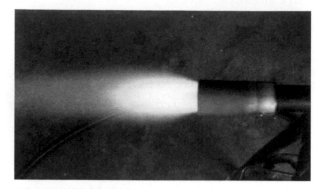

FIGURE 8-21: **Ideal flame** USED WITH PERMISSION FROM RON REIL

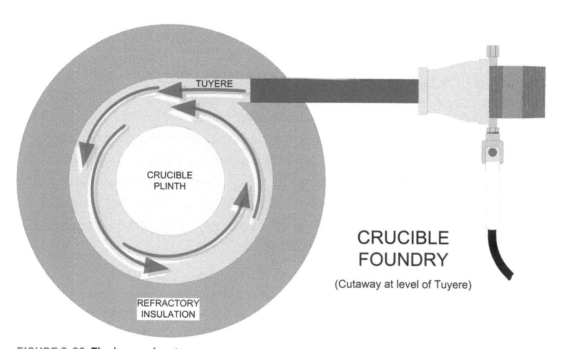

FIGURE 8-22: **The burner in a tuyere**

Operation

Operating the Venturi burner basically follows the same steps for starting and stopping the burner that we undertook in testing. There are a few other factors that depend on how you intend to use the burner.

For use in a crucible foundry, the burner will be inserted tangentially into the wall of the foundry through a hole called a *tuyere* (shown in Figure 8-22). This allows the flame to enter the foundry and swirl around the interior walls, evenly heating the crucible. The nozzle of the burner does not extend into the interior (or it will become excessively hot and add turbulence).

Foundry burners are lit with the burner in place. Don't try to light a burner and insert it into a tuyere. Drop a lit piece of paper into the foundry and slowly open up the regulator. *Do not* wait to drop the paper in after you turn the propane on; you do not want a big fireball to erupt out of the foundry (it's not safe for you, and concussive ignitions are no friend to refractory linings).

Burners mounted in the side of a forge should be lit with an ignition source as close to the nozzle as possible. There are many different forge designs; open, closed, multiple burners, single burners, and so on. Consult the instructions for the forge type you're using to safely ignite it.

Enhancements

Flares

There is a great deal of discussion on the Internet and elsewhere about flares on the end of burners. Flare design can be an exact science, but reasonable results can be obtained with very rough efforts. The most important aspect of a flare is that it must have a 1:12 ratio of expansion. This expansion serves to create an additional degree of vacuum that accelerates and increases the air intake, which in turn allows for a higher volume of propane to be injected, creating a hotter and more stable flame.

The best flares are robust enough to withstand repeated, heavy use. Most of the homebuilt versions are forged out of iron pipe. Forging a flare like this isn't terribly complicated, but you pretty much need a flared burner to create the heat to forge a flare. You can do this by using the burner in a crucible foundry with the foundry used "off-purpose" to heat up the pipe. But I'm also going to describe two basic flares that will allow you to bootstrap yourself into a better flare. These flares don't perform nearly as well as more advanced flares, but they will suffice to get your burner working and usable enough that you can forge something more robust. I cannot recommend enough that you read the pages from Ron Reil's website at www.abana.org. It offers much more information about constructing flares.

Each of the two "bootstrap" flares has advantages and disadvantages. The primary advantage of each is the ease of fabrication with simple tools. Of the two flares presented, the mini-flare produces a cleaner flame, but at low pressure. It's easy to thread on and off the nozzle and toss in a toolbox or pocket. It's durable and should last a long time. The sheet-metal flare produces a much hotter flame at significantly higher pressure. However, it will disintegrate over the course of a number of burns (the number depends on time, temperature, gauge of sheet metal, and other factors). It's relatively easy to make and install, but the flame it produces is rough.

The Mini-Flare

For this flare, we're going to modify a copper ¾" FIP—1" cup adapter by expanding the unthreaded cup end to provide the flare. The amount of flare required to achieve a 1:12 ratio on this short of a fitting is a ¹⁄₁₆" increase in the diameter at the opening of the fitting. This equates to a 2.4° outward angle around the wall of the fitting. (See Figure 8-23.)

The precision with which you create the flare, especially on such a short stretch, is not likely to be very high, but this is okay. The fitting is cheap enough that

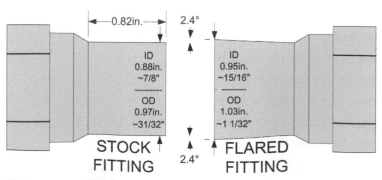

FIGURE 8-23: Mini-flare dimensions

you could try multiple times, or, since the copper is quite soft, you could rework this fitting until it suited your needs better. You could also heat the copper fitting to make it more malleable. You may find that you're able to get it close enough without having to strive for a degree of accuracy that is wasted due to other variances in the system.

The first task is to create a template so that we can determine when the fitting is properly flared. (See Figure 8-24.)

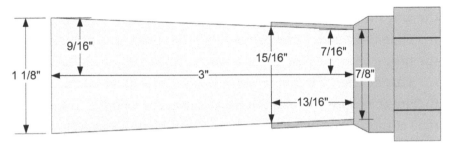

FIGURE 8-24: Mini-flare template

You'll want this to be fairly stiff, so use an index card or some other kind of card stock to cut it out from.

1. Draw a 3″ line, then draw a line perpendicular to it at each end.

2. Mark the left perpendicular line 9/16″ above and below the 3″ line.

3. Mark the right perpendicular line 7/16″ above and below the 3″ line.

4. Connect the mark above the 3″ line on the left to the one on the right.

5. Connect the mark below the 3″ line on the left to the one on the right.

6. Draw a line perpendicular to the 3″ line that is 13/16″ from the right end, which will be the depth marker.

Use the template frequently as you flare the fitting to check progress. When the template fits, without bending, all the way to the 13/16″ depth line, the flaring is complete. We only need to increase the diameter at the end of the fitting 1/16″, so we want to go slowly and check the template often. Ideally, we want to achieve an even angle from the start of the flare to the opening.

Place a ¾″ × 36″ wooden dowel into the ¾″ copper fitting. This fitting should be able to slide smoothly into the copper adapter we are going to flare. This works because, on the cup (unthreaded) side, the fitting has the same outer diameter as ¾″ copper tube. The wooden dowel should fit snugly into the interior of this fitting. It serves to keep the fitting from collapsing. You may need to set the fitting in a vise and tap the wooden dowel to get it to fit all the way in.

Slide the wooden dowel into a ¾″ × 36″ black iron pipe. Thread the fitting onto the pipe as far as you can. (See Figure 8-25.)

This device, I'll call it a *planishing lever*, will serve as our tool to flare the copper adapter. During its use, the wooden dowel may compress, allowing the copper fitting to do so, as well. We want the copper fitting to stay as straight as possible, so you may need to hammer out the dowel from time to time, cut it, or flip it so that a new ¾″ diameter section is at the tip of the fitting. This may require going back to the vise and hammering the dowel into the fitting to straighten it out (you shouldn't need to remove the black iron pipe). If, for some reason, you have a metal ¾″ rod, use it instead to avoid this problem.

Thread the ¾″ copper adapter that will be flared into the 8″ section of black iron pipe (P1). This piece of pipe will serve as our mandrel (handle) for the adapter. Slide the fitting on the end of the planishing lever into the adapter. Now, find some solid height differential of 4–10″; I'll refer to this as the frame.

By *solid height differential* I mean stairs, a curb, or a floor and a 4 × 4 beam—essentially any set of rigid surfaces where we can rest the mandrel on edge at the top and the end of the planishing lever at the bottom. (See Figure 8-26 on the next page.)

Once this is set up, we'll do two things: put pressure downward on the copper fitting and adapter and roll the whole setup along

FIGURE 8-25: Dowel, fitting, and pipe

the frame so that the pressure is evenly distributed around the full circumference of the adapter.

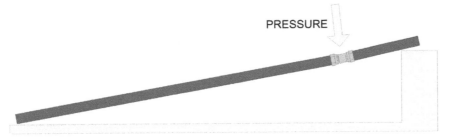

FIGURE 8-26: The flaring frame

Remember, we're only trying to achieve a small angle in the wall of the adapter; 2.4° is our target. Go slow and check the template (and the state of the dowel in the planishing lever) frequently. You can see how I did it in Figure 8-27.

The end of the adapter is more prone to flaring than the base where it's attached to the threaded section. This is no surprise, but a major reason that you want to be sure that the fitting on the planishing lever doesn't get rounded at the tip by the dowel compressing. All in all, it's a relatively easy and cheap way to build a starter flare! (See Figure 8-28.)

Sheet-Metal Flare

This flare is made from light-gauge sheet metal. You'll only need a small piece of sheet metal, 4½″ × 5″, but it's good to have extra for experimentation. (See Figure 8-29.)

FIGURE 8-28: The completed mini-flare

FIGURE 8-27: Flaring the adapter

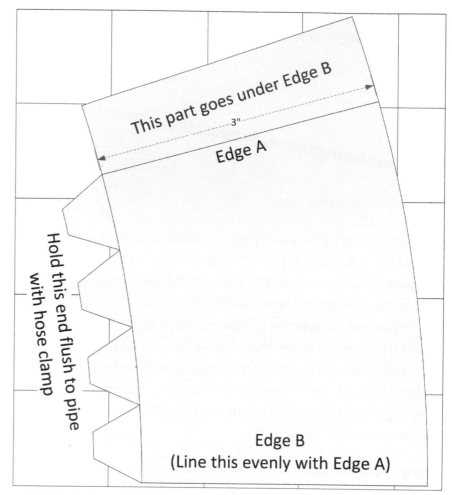

FIGURE 8-29: Sheet-metal flare template

You can copy this image, cut it out, adhere it to the sheet metal, and use it as a guide to cut the metal with snips. If you're unable to copy it, you can re-create the template with a few simple practical geometry steps.

1. Get some string, some tape, and a fine-tip marker.

2. Tape one end of the string at the corner of the sheet metal so that it is at the very edge of the sheet. This will need to be at least 15″ away from the metal you will be marking. If the metal is too small for this, we can create a "virtual" larger sheet. Secure the metal to a surface, like a table, with tape. Hold a straight edge against the left edge of the metal so that it extends down the table about 20″. This

is the left edge of our virtual sheet. Tape the string to the table on the virtual edge at least 15″ below the top of the piece to be cut.

3. Tape the other end of the string to the marker so that it will make a circle with a radius of 11½″.

4. Draw an arc onto the metal (with that 11½″ radius) starting at the edge of the sheet. The arc should be about 5″ long (length isn't critical).

5. Retape the marker on the string so that you can draw an arc with a radius of 12″, and draw another arc above the first one.

6. Retape the marker on the string so that you can draw an arc with a radius of 15″, and draw a third arc above the second one.

7. On the 12″ arc, use a ruler and draw a 3¼″ line from the edge of the sheet where the arc starts, and ending where it touches the arc.

8. Draw another line on the 12″ arc that is 4″ long, starting at the edge of the sheet where the arc starts and ending so that it touches the arc.

9. Draw a line on the 15″ arc that is 4^1/$_{16}$″ long, starting at the edge of the sheet where the arc starts, and ending so that it touches the arc.

10. Draw another line on the 15″ arc that is 5″, starting at the edge of the sheet where the arc starts and ending so that it touches the arc.

11. Connect the short lines (3¼″ and 4^1/$_{16}$″) with a line.

12. Connect the long lines (4″ and 5″) with a line.

13. Draw six diagonal lines between the 11½″ arc and the 12″ arc to create four evenly spaced tabs (see Figure 8-30).

Your markings should look like those in Figure 8-30.

I marked the area to cut with red lines (they'll look like all the other lines on your sheet). You can cut the sheet with heavy scissors or tin snips.

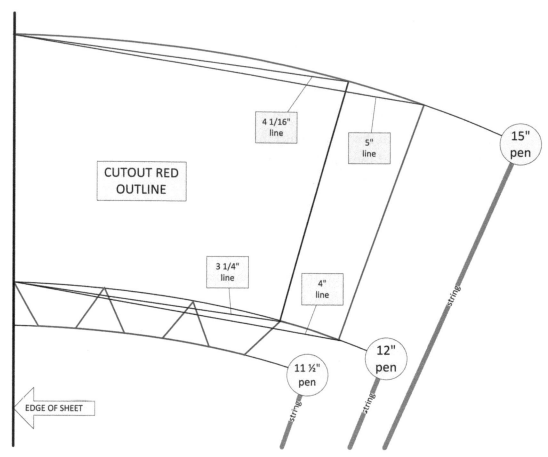

FIGURE 8-30: Sheet-metal flare layout

Once you have a good flare shape cut out, hold the overlap edge against the 8″ section of ¾″ pipe and start rolling. The goal at this stage is not to roll it tightly, but to just gently create an even curve across the length of the flare. Roll until edge B lines up against edge A cleanly (the large end of the flare will be an even circle).

Place the 1½″ hose clamp over the tabs at the narrow end of the flare. Position the clamp with a small overlap (no more than 1/16″) above the tabs on the flare. Slide the flare and hose clamp onto the end of the 8″ pipe and tighten the hose clamp as tight as it will get. If you see bends in the flare that create daylight at the narrow end, you can carefully tap them down with a hammer or remove the flare and continue to hand roll and bend them out.

Safety Note

Most easily cut sheet metal is galvanized steel. This means that the steel is coated with zinc. When heated sufficiently, zinc can burn or give off white fumes that are harmful. The acceptable limit is considered to be 5 milligrams per cubic meter (mg/m³) of air averaged over a 10-hour day in a 40-hour week, or 15 mg/m³ over 15 minutes.

There is a lot of misinformation about zinc fume exposure. Short-term overexposure (higher than what is defined above) can result in feeling like you have the flu. Headache, fever, chills, aches, nausea, and tiredness are all possible symptoms. This can last 6–24 hours. The Occupational Health and Safety Administration (OSHA) describe no known long-term exposure hazards. Unless there are complicating factors, it isn't fatal, and most people who have attacks build up some degree of immunity over time (but quickly lose it if not around the fumes). Nevertheless, if you work where you are going to breathe zinc fumes, you should wear a respirator.

In well-ventilated situations, the exposure limits are higher than most people will encounter dealing with small galvanized parts. It would be almost impossible for a 5″ square piece of galvanized sheet metal, heated outdoors in a well-ventilated area, to reach the exposure limits for someone nearby. However, for your safety, do not use galvanized sheet metal for your flare unless you are fully aware of the risks and how to mitigate them.

You can improve the output if you use metal ducting tape to wrap the back end around the pipe. The tape will keep air from getting in. You can even use the tape instead of the hose clamp. The flare should not get extremely hot unless air is entering from the back or sides and the flame is inside the flare, so the tape should be okay. A strip along the seam will also help.

With this flare, the burner can do an open-air burn with up to 10 psi. However, this flare will not last through very many burns before it will eventually corrode. But you can get quite a few flares out of most pieces of sheet metal, so you can experiment with this design and use the burner to forge a proper flare out of more robust material.

The flare projects 3″ beyond the end of the pipe when complete. (See Figure 8-31.)

FIGURE 8-31: The completed sheet-metal flare

Chokes

The axial choke we built in this project is easy and effective, but prone to getting knocked out of alignment. For a burner that is going to be in heavy use, a more robust axial choke is a big advantage. With access to welding gear, or even silver soldering or brazing and some careful cutting and grinding, it is possible to construct a much better axial choke. (See Figure 8-32.)

Other choke configurations are possible, but the axial design is the best way to maintain laminar flow of air into the burner.

Flow

The burner is a fundamentally simple device. The Venturi principle allows a small amount of propane to draw in a large amount of air. Improving the output and performance of the burner

FIGURE 8-32: Upscale axial choke
Used with permission from Ron Reil

basically means that you're working on one of four things (see Figure 8-33):

1. The smooth laminar intake of air prior to mixing

2. The insertion of propane via the jet

3. The turbulent mixing of the propane and air

4. The expansion at the flare to drop the pressure and increase the draw of air into the system

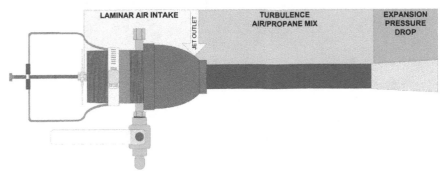

FIGURE 8-33: **Zones of action in the burner**

The best solutions for each of these take advantage of an advanced understanding of fluid dynamics. Aspects of some are literally rocket science. That doesn't mean that you can't build and improve on a fantastic burner with hand tools and a basic understanding of the parts; it just means that there is always room for improvement.

When your burner would benefit from enhancement, or needs troubleshooting, break the problem down into one (or more) of these four areas. Take advantage of Ron Reil's website and the wealth of other information on the Internet. Perhaps, like Faraday finding the cores of chemistry and physics in a candle flame, you'll find a gateway to a deeper understanding of engineering and physics in your simple burner.

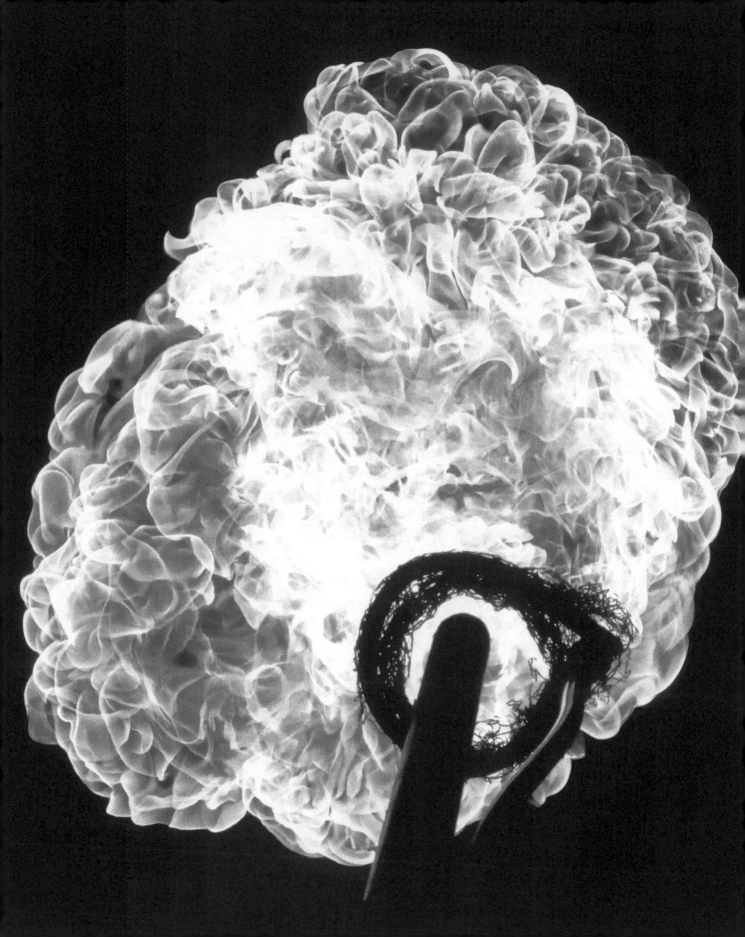

9

The Boosh

EVEN THOUGH I HAVE a soft spot in my heart for all the projects in this book, the boosh (aka *poofer*, *flame cannon*) is the one this book is really about. If you've jumped ahead to read this chapter first, please know that the other chapters and projects provide essential skill and safety information. This chapter cannot stand alone; without reading those other chapters, you'll be missing some key information.

Flame effects provide a powerful visceral experience. They engage nearly all the senses (hopefully not taste) and push people's buttons. There just really isn't any other experience like setting off a huge ball of fire in the sky. The goal of this book is to learn how to do this without endangering yourself and others. I want you to be able to enjoy this experience, but I have to repeat some commonsense warnings.

Flame effects that produce a large jet or ball of fire can be produced with a wide range of fuels. The challenges are dispersal, air/fuel ratio, ignition, and general safety. Propane provides a number of advantages when taking on these challenges. Since propane is in common use, there is a huge infrastructure set up to make it safely available almost everywhere. Propane is stored, transported, and sold as a liquid under pressure. Commonly available regulators step this down to usable levels. Simply venting propane vapor handles dispersal. In most cases, that dispersal does a reasonable job of mixing with sufficient air by virtue of the velocity with which it exits and expands.

With big advantages in dispersal, air/fuel ratio, and many aspects of general safety (as opposed to safety issues surrounding storing and handling toxic or more corrosive fuels) our challenges are the safety aspects specific to the device, and ignition. We'll concentrate on both of these topics. The safety aspects are

Safety Note

A flame effect, as described in this book, must only be operated or directly supervised by a responsible adult. Construction of effects, such as the one described below, must follow all safety guidelines and directions. Operating or constructing this effect in an unsafe manner may cause injury or death for which the author and publisher are not responsible.

You must follow all laws pertaining to the operation of an effect like the one described. Do not ignite a flame effect in jurisdictions where it is not allowed. Contact the local authority having jurisdiction (AHJ) or read the appropriate statutes and codes to understand the laws that pertain to the area you are in. You are responsible for knowing the law; ignorance is no excuse.

Do not operate a flame effect with bystanders closer than 20′ (6 m) to the flame produced. Restrict access to the effect and its controls, and never leave it unattended unless it has been completely secured. Follow all required laws, codes, and standards for having fire extinguishers ready and on hand. Understand the requirements for having individuals trained in their use and ready to help as fire wardens. Control access to fuel sources and keep them at least 10′ (3 m) from the effect.

Leak-test your effect before every use. Make sure your propane hoses are not placed where someone will step on them. Be sure to have a "kill switch" that will fully disarm the effect in the case of an emergency. Always create a plan for setup, operation, and teardown that describes the effect and how it will be used safely.

These requirements may sound like a lot of frustrating constraints. But like many things that can be safe to use, there are standards that must be followed for them to be so. Setting things on fire is no challenge, but doing so in a repeatable, exciting, and safe manner is what we're after in this book. Stay safe and you'll be able to get your effects permitted and fire them up at Maker Faires, burn events, and other places knowing you're doing it right.

considered throughout the book, and we'll be covering ignition in this and subsequent chapters in depth.

If you've driven past an oilfield or refinery at night, you may have seen them flaring gas. (See Figure 9-1.)

This is an industrial-scale, continuous-operation, flame effect built with extremely expensive technology. Few entertainment flame effects can rival the amazing power of a big gas flare. But most effect operators want to be able to fire the effect on-demand rather than run it continuously. The design goal for most flame effects is to control the discharge of fuel and ignite it reliably and on-demand.

Let's look closely at the key requirements; fuel, control, and ignition. Variations in how they are approached constitute almost all the differences between a basic flame effect like the boosh project in this book and a giant stadium concert effect.

Fuel

This book is focused on propane vapor. Exciting flame effects can be built with a wide variety of other fuels, each of which presents unique challenges. For this text, I stick to propane vapor, although I briefly discuss the use of liquid propane in Chapter 13, "Beyond Vapor."

Great effects can be constructed that use very small amounts of vapor. However, for a variety of reasons, there is something of an arms race that seems to kick in when people start building flame effects. The goal frequently becomes bigger fireballs created by releasing ever larger amounts of vapor. In most cases, it's desirable to release the vapor in as short an amount of time as possible. People want to create a big fireball instead of a long-running jet. Since the cylinder (always with a regulator!) can only release as much vapor as is present in its head space (see

FIGURE 9-1: Gas flaring from a drilling platform
PUBLIC DOMAIN IMAGE, WIKIMEDIA COMMONS

Chapter 1, "Understanding Propane," if you don't remember why) most designers create a reservoir of vapor somewhere in the system. This is referred to as an *accumulator*. We discussed what can be used as an appropriate accumulator in Chapter 2, "Equipment and Parts."

One of the primary ways to get a bigger effect is to increase the capacity of the accumulator. Code and common sense tell us that an accumulator shouldn't be larger than what will be used for the intended effect, so we don't want the accumulator to be a primary fuel source, just the reservoir needed to fire the effect.

The primary source, the propane cylinder, is another aspect of fuel that differs between propane vapor flame effects. Not that there are different types of propane; the main issue is whether the fuel source delivers enough vapor to recharge the accumulator in a reasonable time. This is problematic since, as we discussed in Chapter 1, the more vapor you draw from a propane cylinder, the colder it gets and the less pressure it can produce. It's incredibly common to see 3–4" (75–100 mm) of ice on a cylinder after a booshing session. This is usually accompanied by a sad look on the operator's face as tiny little burps of 5 psi vapor emerge from the effect.

The size of the cylinder doesn't make a tremendous difference, since all cylinders at the same temperature are at the same pressure, regardless of size. However, the thermal inertia of a mass of 80 lbs of liquid propane will take longer to freeze than a mass of 4 lbs so bigger or multiple cylinders can be useful.

Many large or commercial flame effects will use warm water baths or specially made cylinder heating blankets to keep the cylinder from freezing up. Trying to heat a cylinder filled with a mixture of propane liquid and vapor is extremely dangerous. *Do not* attempt to do so unless you are using purpose-made parts and have many years of experience. Raising the cylinder pressure too high is a recipe for disaster.

The last major design difference having to do with fuel is the size of the effect valve and the passageway from accumulator to vent. When attempting to release a large volume of vapor in a short time, it makes sense that the diameter of the outlet path is a limiting factor. Even the difference between a ½″ and a ¾″ passage is noticeable. Effects using 2″ and 3″ valves can produce a phenomenal fireball that will stun you with its concussive force.

Get comfortable building effects with smaller valves before attempting to design large-diameter systems. The larger valves are painfully expensive and really require a system with everything else tuned well to support it. They also consume propane faster, take longer to charge, freeze cylinders quicker, and are generally a lot more challenging and expensive to operate.

One side note about sizing the valve and passageway from accumulator to vent: if the accumulator has a ¾″ threaded opening, you're wasting the 2″ valve since the ¾″ opening acts as a resistor in the path. Either a custom accumulator with a 2″ opening or multiple accumulators are needed to really take advantage of larger components. The implications of this are worth considering for the rest of the system, as well. The narrowest passage in your system, between the regulator and accumulator, defines the volume that can charge the accumulator. There's no advantage in using 1″ pipe for parts in the charging path if the outlet from the accumulator is ¼″. The only exception to this is that long runs of hose can serve to drop pressure, so it's worth over-sizing them to a degree to avoid constraints.

Control

The effect valve can be controlled by two main methods. It can be manually controlled with a whistle valve or with an electromechanical solenoid controlled by a switch or a computer. Manual whistle valves are rarely seen and shockingly expensive, running close to $140 USD for a ¾″ valve as I write this. A ¾″ 12V DC solenoid is about $30 USD for comparison. I'll stick to solenoid valves in this text.

Control is important because, if the effect is operated by a "show control system," the classification of the effect changes.

Almost all flame effects built by amateurs are classified as Group III effects: "An attended, temporarily installed flame effect for a specific production with limited operation and fixed time for removal." Higher group ratings that involve show control system–operated effects have more stringent requirements.

Having a microcontroller or other computer fire the solenoid valve is not, by itself, a show control system. According to John Huntington in *Show Networks and Control Systems* (Zircon Designs Press, 2012): "The key is that you don't have a *show control system* unless control for *more than one* production element is linked *together*" (Italics in the original). Show control systems often include sensors to recognize whether performers are in safe locations (or they don't fire the effect), coordinate with music and lighting, and run a whole course of coordinated actions once started.

Once you're looking for bigger challenges for your flame effect skills, you may want to look into the professional world of show control. In this book, we will build our boosh with manual switched control of the solenoid and, in a later chapter, build an Arduino shield to fire it in patterns. It's a great place to start.

Ignition

Amateurs talk about tactics, but professionals study logistics.
GEN. ROBERT H. BARROW, USMC (COMMANDANT OF THE MARINE CORPS)

With flame effects, professionals focus on ignition. Building an effect and inadvertently filling the surrounding air with unignited propane vapor is a menace. Getting it to ignite every time, in wind, under different pressures, on-demand is the task that ends up being really challenging.

Pilot lights are a tried and true approach to main effect ignition. Some are always on, and some are lit just prior to the main effect. Some are manually lit and some remotely lit. High-end flame effects will occasionally do away with pilots altogether and construct a direct ignition source.

We will build the boosh in this chapter with the simplest and most reliable of ignition sources: the continuous low-pressure pilot. It's a very easy design to be successful with. In the chapter on pilots we'll discuss and build a remotely lit pilot that you may want to build as an upgrade.

Parts

The schematic diagram for the boosh has quite a few more components than our previous projects. (See Figure 9-2.)

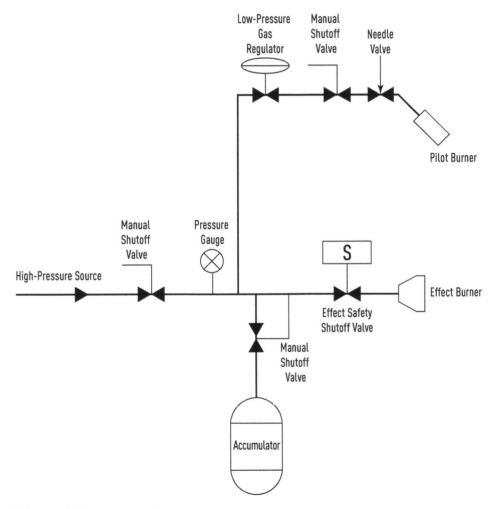

FIGURE 9-2: **Boosh schematic**

Boosh Parts:

REF	ITEM	QTY
E1	High-pressure propane hose 10′ ⅜″ FFL × ⅜″ FFL *	1
E2	Flare adapter ⅜″ MFL × ⅜″ MFL *	1
F1	Black iron cross ¾″	1
F2	Black iron pipe nipple ¾″ × 2″	2
F3	Black iron pipe nipple ¾″ × 6″	1
F4	Brass flare plug ⅜ ″	1
F6	Brass adapter ¼″ MIP × ⅜″ MFL	2
F7	Brass flare nut ⅜″	2
F8	Brass bushing ¾″ MIP × ½″ FIP	1
F9	Brass bushing ¾″ MIP × ¼″ FIP	1
F10	Brass reducing nipple ¼″ MIP × 1/8″ MIP **	1
F11	Black iron tee ½″	1
F12	Brass bushing ½″ MIP × ¼″ FIP	1
F13	Black iron pipe nipple ½″ × 2″	2
M1	Gauge 0–100 psi ¼″ MIP	1
P1	Black iron pipe ¾″ × 6′	1
P2	Copper propane-rated refrigeration tube ⅜″ × 6′	1
P3	Accumulator (decommissioned 20 lb propane cylinder)	1
S1	Copper scouring pad	1
S2	Bolt ¼″ × ½″	2
S3	Lock washer ¼″	2
S4	Nut ¼″	2
S5	Hose clamp ½″	2
S6	Conduit hanger ¾″	2
S7	Safety (or lock) wire (.03″–.05″) × 12″	1
V1	Gas-rated solenoid valve 12 V DC ¾″ FIP × ¾″ FIP	1
V2	Needle valve ¼″ FIP × ¼″ MIP	1
V3	Ball valve ¼″ FIP × 1/4″ FIP	1

REF	ITEM	QTY
V4	Ball valve $1/2''$ FIP $\times$ $3/8''$ MFL	1
V5	Ball valve $3/4''$ FIP $\times$ $3/4''$ FIP	1
V6	Low-pressure regulator with hose $1/8''$ FIP $\times$ $3/8''$ FFL **	1

* The requirement is to extend the high-pressure source's hose so that you have at least 15' (4.5 m) of hose between the fuel and the effect. Any combination of hose lengths (original, plus additional) that gives you at least that minimum length is acceptable. The additional hose must come with a $3/8''$ female swivel flare fitting on at least one end. If it has $3/8''$ female swivel flare fittings on both ends, you will need the male/male flare adapter ($3/8''$ MFL $\times$ $3/8''$ MFL) described above to connect it to the existing hose. If the additional hose comes with a threaded fitting ($1/4''$ MIP, $3/8''$ MIP, etc.) on the other end from the female swivel flare, you will need a $3/8''$ MFL $\times$ [appropriate size] FIP adapter to connect it to the existing hose.

** Your low-pressure regulator may come with a $1/4''$ FIP inlet (which you may not discover until you remove the cylinder fitting). If so, replace part F10 with a $1/4''$ MIP $\times$ $1/4''$ MIP $11/2''$ brass nipple.

Control Box Parts:

REF	ITEM	QTY
C1	Lighted red arcade momentary contact switch 100 mm	1
C2	Key switch	1
C3	Emergency stop "mushroom" switch	1
C4	Cookie tin $6'' \times 8'' \times 2''$	1
C5	Two-wire lamp cord 18' (could be two 12' cords)	1
C6	Grommets $9/32''$	2
C7	Wire nuts (appropriate for 16-gauge wire)	4
C8	Two-pin trailer connector jumper	4
C9	Replacement 12V cigarette lighter plug	1
C10	Heat shrink tubing (various sizes)	1

Base Parts:

REF	ITEM	QTY
Z1	Plywood $4' \times 4'$ ($1/4''$ or greater thickness)	1
Z2	Beam clamp $1'' \times 11/4''$	4
Z3	Bolts $1/4'' \times 3/4''$	4

The block diagram is large enough that I've broken it into two parts. The first presents the central "plumbing tree" that includes everything except the pilot and main vent. The second details the pilot system and the main vent pipe. (See Figures 9-3 and 9-4.)

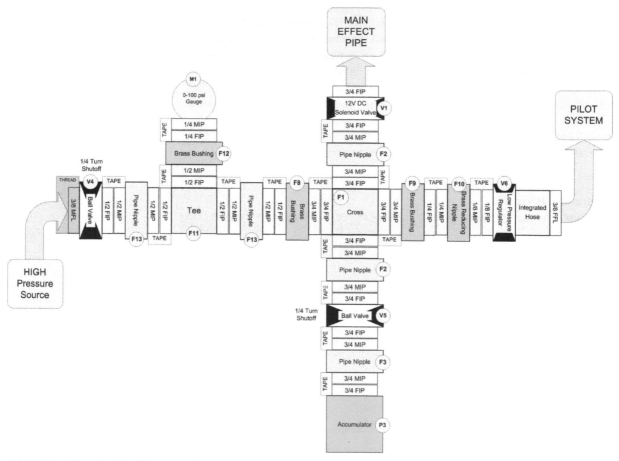

FIGURE 9-3: Boosh plumbing tree block diagram

Tools

The new tool that we'll use in this chapter is a flaring tool. This tool consists of two parts: the cone and the die. These are frequently sold as a kit with a tubing cutter and possibly tubing benders. There are expensive heavy-duty flaring tools that are a joy to use, but the standard flaring tool sold at hardware or tool stores will work great. We'll only be doing single flares, so you don't need a double flaring kit.

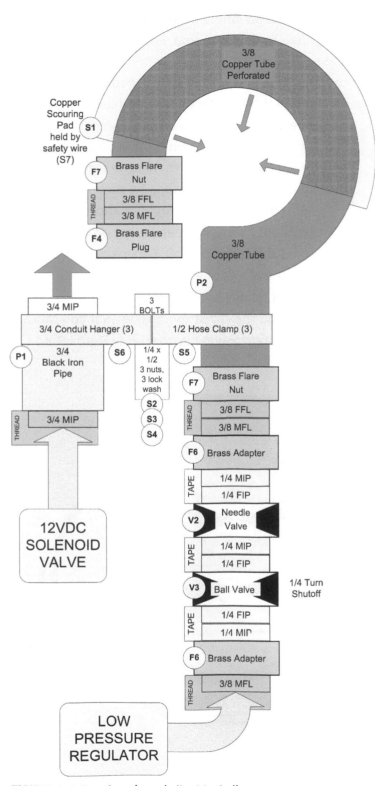

FIGURE 9-4: Boosh main and pilot block diagram

Most double flare kits will do single flares as well, but unless you're doing things like brake lines, you won't need the double flare capability. (See Figure 9-5.)

The other tool that comes in very handy is a multimeter. The cheapest multimeter will work perfectly for our needs. The only feature that I'd miss on the cheapest meters is the audible tone on the continuity setting. With this feature, if there is an electrical path between the leads, the meter beeps. This allows you to search for shorts or breaks without having to keep an eye on the meter. Useful, but not critical. Expensive multimeters have lots of great features, but you won't regret having some cheap ones around (to leave in the car or toolbox) even if you buy a more expensive one later. (See Figure 9-6.)

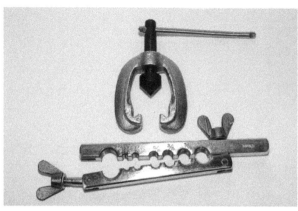

FIGURE 9-5: **Flaring tool**

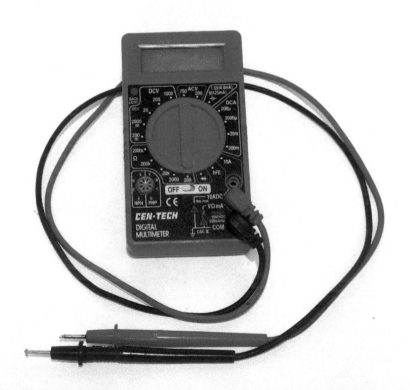

FIGURE 9-6: **Multimeter**

Construction

1. Assemble the Central Plumbing Tree

When breaking down a device like this for storage or transport, it's useful to have predefined places that easily assemble and disassemble. The plumbing tree contains the majority of the components in the boosh and is intended to remain together as a unit. The other components can be attached to and detached from it fairly quickly and with simple tools. It's undesirable to have a taped joint be a break point; flanged fittings are ideal for this since they are quick and easy to thread and unthread.

The main vent pipe is the exception to the flanged fitting break-point guideline. However, since it is an unpressurized section of the unit, connected above the 12V DC solenoid, it can be threaded hand-tight without tape, making the fitting easy to put together and take apart.

The black iron pipe "cross" is the center of this unit. These can be sometimes difficult to find locally. If you don't want to buy one online, you can substitute two black iron tee fittings joined with a short pipe nipple to get the same number of inlets and outlets as the cross. If you do use two tees, make sure to tape the joints for the pipe nipple and to align the parts so that they aim in the desired directions. (See Figure 9-7.)

I'll describe the subunits leading out from the cross as *branches*. There are four branches: accumulator, gas inlet, pilot, and main. (See Figure 9-8.)

FIGURE 9-7: **Two tees as a substitute for a cross**

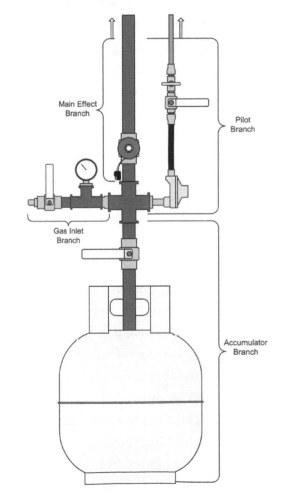

FIGURE 9-8: **Boosh branch layout**

Accumulator Branch

I will recommend, at a later stage, that you paint the effect, including the accumulator (see "5. Final Assembly for Testing"). Some people will wonder why I didn't tell them to paint the accumulator before they plumbed it into the system. If that makes more sense to you than plumbing it in and then painting, please read that section and paint at will. I recommend putting a ¾" MIP plug into the threaded hole on top if you paint at this stage.

Chapter 2 describes how to acquire an accumulator from a propane distributor. We will be using a 20 lb (aka 5 gallon) decommissioned propane cylinder (P3) as our accumulator. These cylinders have a standard ¾" FIP connection once the valve is removed. Put Teflon tape on the threads of the 6" length of ¾" black iron pipe (F3) and thread it hand-tight into the cylinder opening. Tape the top end of the pipe and thread a ¾" quarter-turn ball valve (V5) onto it. Using a crescent wrench on the valve, tighten the pipe and valve together.

This set of fittings needs to be strongly tightened. They will be holding pressure on the main vent path. Since this is also the mechanical base for the effect, the fittings are not only providing pressure protection, they will also be holding up the entire effect. Be careful with all the plumbing, but be sure to start with a solid foundation.

We could have tightened the pipe into the cylinder with a pipe wrench. That's a great way to do it if you want to. However, crescent wrenches are my preferred tool because they should have a positive lock on the hex fitting. Pipe wrenches can slip and leave gnarly marks on your pipe. Don't get me wrong; pipe wrenches are frequently the ideal tool to use, but when you can accomplish the same task with a crescent wrench, I prefer to do so.

Make sure, as you finish tightening the ball valve and pipe into the cylinder, that you end up with the ball valve handle pointing the direction you want. I always like mine to face the opening in the cylinder handle, but you may have a different preference. The way this handle faces is going to be the front of the effect, so work to get it where you want. This should mean tightening more, not less, to get the fitting in your desired position, so you need to think about where you want it to be well before you get to the point where you can't turn any more.

Once the ball valve and 6" pipe are in place, tape the threads on both ends of a ¾" × 2" pipe nipple (F2). Thread it into the top of the valve and thread the ¾" black iron cross fitting (F1) onto it. Once again, tighten both the nipple and the cross together. You'll want to hold the ball valve in place with a crescent wrench while you do this so that the torque of the cross and nipple turning don't move it out of place. You can also thread a segment of pipe into one of the horizontal cross arms to turn it.

This completes the accumulator branch and, with the introduction of the cross, sets us up for the other branches. (See Figure 9-9.)

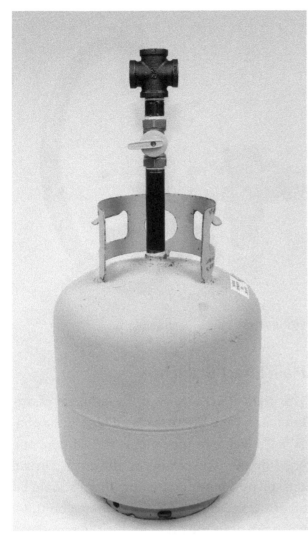

FIGURE 9-9: Accumulator branch

so that the central opening points upward when complete.

Tape both ends of the ½" × 2" pipe nipple and thread one end into the ½" tee. Thread the ½" quarter-turn ball valve (V4) onto the nipple and tighten both together into the tee so that the handle of the valve faces forward. You will want to hold a wrench on the tee to make sure it doesn't move as you tighten the valve and nipple into it.

Tape the ½" MIP × ¼" FIP brass bushing (F12) and thread it tightly into the top of the ½" tee. Tape the ¼" MIP threads on the gauge (M1) and thread it into the bushing. Tighten the gauge so that it faces forward.

This completes the gas inlet branch. (See Figure 9-10.)

Pilot Branch

The pilot branch will include everything up to the pilot vent tube. We start by taping the external threads on the ¾" MIP × ¼" FIP brass reducing bushing (F9). Thread this into the right arm of the cross and tighten. Tape both ends of the ¼" MIP × ⅛" brass reducing

Gas Inlet Branch

Construction of the gas inlet branch begins at the cross and works outward. Tape the ¾" threads on the ¾" MIP × ½" FIP brass bushing (F8) and tighten it into the left arm of the cross (that's arbitrary; you can put it on either side, but I want to match the diagrams). Tape the ½" threads on the reducing nipple and tighten the ½" tee (F11) onto it

FIGURE 9-10: Gas inlet branch

nipple (F10) and thread one end into the brass reducing bushing.

Thread the ⅛″ FIP input on the low-pressure (LP) regulator (V6) onto the end of the nipple and tighten so that the regulator's integrated hose is pointing up when complete.

Your LP regulator may have come with a cylinder fitting threaded into the ⅛″ input. If so, we'll need to remove this. We can do so with an ¹¹⁄₁₆″ socket wrench.

It's also possible that your LP regulator has a ¼″ FIP inlet instead of a ⅛″ FIP inlet. If so, swap the ¼″ MIP × ⅛″ MIP reducing nipple for a ¼″ MIP × ¼″ MIP nipple.

The next part of the pilot branch is almost independent. It consists of the needle valve for adjusting the pilot flame and the pilot cutoff valve. Start by taping the male threads on the needle valve (V2) and threading it into the ¼″ quarter-turn ball valve (V3). Be careful that the ball valve is placed so that, when the handle is in the open position, it opens away from the needle valve and the two don't obstruct one another. Tighten these valves so that the controls for both face the same direction.

Tape the ¼″ threads on the ¼″ MIP × ⅜″ MFL adapter (F6) and thread it into the ball valve. Do the same with the second ¼″ MIP × ⅜″ MFL adapter and thread it into the needle valve. Tighten appropriately. Tighten the ⅜″ FFL swivel fitting on the end of the low-pressure regulator's integrated hose onto the ⅜″ MFL fitting attached to the pilot cutoff valve. (See Figure 9-11.)

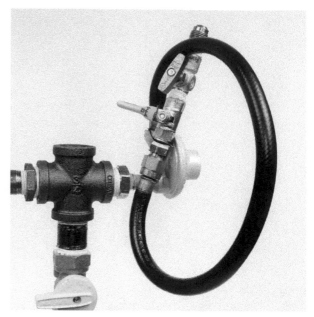

FIGURE 9-11: Pilot branch and valves

Main Effect Branch

The main effect branch is the simplest of the branches. Tape the threads on both ends of a ¾″ × 2″ black iron pipe nipple (F2). Thread one end into the top opening on the cross. Thread the solenoid's input opening (there should be an arrow indicating direction of flow on the solenoid) onto the other end of the nipple.

Tighten the solenoid and nipple firmly so that the coil faces forward (or whichever direction you prefer) when you're done. You can hold the gas input branch while you tighten to keep the cross from turning out of the desired position.

When the boosh is not in use, I remove the ¾″ × 6′ main effect vent pipe (P1) from the top of the solenoid. When it's time to assemble the effect, this pipe can be threaded by hand into the solenoid outlet. It does not need to be taped since it does not hold pressure. We will be mounting the pilot vent pipe to the main effect vent pipe, so they will attach as a unit.

This completes the main effect branch and the plumbing tree. (See Figure 9-12.)

2. Fabricate the Pilot Vent

This pilot is a manually lit, continuous operation design. By *manually lit*, I mean that you have to turn on the gas and light it with a

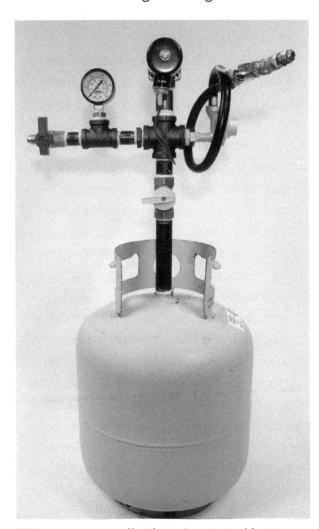

FIGURE 9-12: **Main effect branch and plumbing tree**

fireplace lighter or a torch of some kind (we'll discuss options for this later). By *continuous operation*, I mean that the pilot remains lit for the entire period that you want to operate the effect. With a low-pressure regulator and a needle valve, the pilot uses a negligible amount of propane (certainly compared to the boosh itself) so it's not a significant drain on the main supply.

There's a certain appeal to the continuous pilot. Since most people tend to fire booshes at night for the best visual effect, the pilot serves as a cozy flambeau lighting the sky and stating your intentions to the surrounding environment. It's also very robust and reliable. We'll create a ring of flame around the main vent mouth so that no matter which direction the wind is blowing, the main effect will ignite. This can be a big problem with partial surround or vertical pipe pilots.

Lastly, one of the National Fire Protection Association (NFPA) requirements has to do with visual confirmation that the pilot is lit. A flaming ring flambeau at the top of your boosh is an excellent indicator that you are armed and ready to go!

Cut the Pilot Tube

Cut a 6′ length of the copper refrigeration tube (P2) with the tubing cutter. You may have to use a tape measure and measure the coiled tube in short increments to effectively measure it out. It's okay if you're off by even as much as 6″ (15 cm) since we have the low-pressure hose to help us position it. Ream the tube ends completely so that there is no lip. Gently uncoil the tube so that it becomes reasonably straight.

Bend the Pilot Tube

Make a mark 12″ from one end of the tube. Slide a ⅜″ tube bending spring (like we used in the portable fire pit project) over that end of the tube until it is about 1″ past the mark.

Gently bend the tube in a 90° angle. You ought to be able to bend it with your hands. It doesn't have to be a terribly sharp bend. Slide the spring most of the way off of the bend and then start bending the short arm of the tube into a ring perpendicular to the long section of the tube. It may help to bend against your knee or some rigid object like the edge of a door. You don't want to bend too hard; do it in small increments.

Continue bending the tube around and incrementally slide the tube off as you move the portion being bent closer to the end. The last 2–3″ will be very difficult to bend, but this is okay, because we need to leave a straight segment at the very end that we can flare. Slide the bending spring all the way off the tube. If it gets stuck, twist it in the direction that would "unscrew" it.

It's okay if the copper isn't bent in a perfect circle. We just need it to more or less ring the mouth of the main effect pipe. We'll be covering it in a wind-resistant scouring pad, so it won't be very visible in any case. (See Figure 9-13.)

Flare the Pilot Tube

We need to flare both ends of the tube—one end simply to cap it, and the other end so

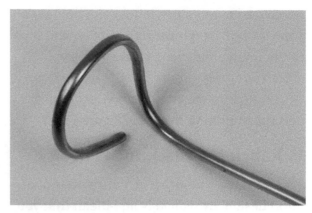

FIGURE 9-13: Bent pilot ring

that it will connect to the needle valve. We'll start with the capped end.

Place one of the ⅜″ flare nuts (F7) over the end of the tube. Clamp the end of the tube in the die with the appropriate amount protruding (see the "Single Flaring Copper Tube" sidebar to determine how much). You will have to gently bend the ring to open it enough to do the flaring on the end, so be careful to only bend it out as far as necessary so that you can bend it back after we've flared it. The safest way to avoid crimping the tube is to put the bending spring back on for this bend.

Attach the flaring cone per the sidebar instructions and flare the end. Remove the cone and die from the tube. Slide the flare nut up to the flared end and insert a ⅜″ flare plug (F4). Tighten with wrenches on both the nut and plug. Bend the pilot ring gently back and close the ring. If you accidentally crimp the tube while bending, adjust a pair of vise grips to the normal width of the tube and clamp them to the ends of the crimp. This will squeeze the tube mostly back into round.

Single Flaring Copper Tube

Flare fittings provide a mechanical alternative to brazing for joining copper tube to other parts of the plumbing system. There are many standards and types of flaring: 45°, 37°, single, double, bubble, and so on. For the purposes that we are going to use it, we'll stick to the simplest: a 45° single flare.

To create the female flared end on the copper tube, you'll need a flaring tool as described in the "Tools" section above. This tool consists of two parts: the cone and the die. The die is sometimes referred to as the anvil. The die splits into two parts and clamps around the outside of the tube to be flared.

The first requirement is to have a square end on the tube to be flared. This is best achieved by using the tubing cutter introduced in previous chapters. Once the tube has been cut, it is essential to ream out the tube end so that none of the lip bending inward remains.

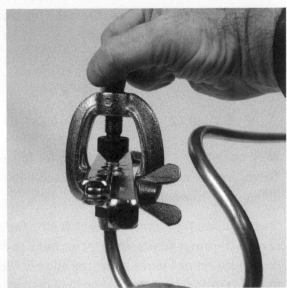

You will practice your expletives if you forget the next step. Before you mount the tube in the die, put the flaring nut onto the tube with the large opening toward the end. (See Figure 9-14.)

Open up the die by loosening the wing nuts, one of which will swivel out of the way to allow the die a larger opening. Slide the end of the tube into the appropriately sized opening. The next step depends on which tool you purchased.

FIGURE 9-14: Include the flare nut before flaring.

We need to have the correct amount of flared tubing bent outward by the cone. *The Copper Tube Handbook*, produced by the Copper Development Association Inc. (copper.org) states:

> *Adjust the height of the tube in the opening in accordance with the tool manufacturer's instructions, to achieve sufficient length of the flare.*

My tool is so old that the instructions are long gone. Looking at instructions provided with new tools, you'll find that some recommend that you mount the tube with the end flush to the die. Some recommend a ⅛" extension beyond the die for all sizes of tube. Others make more cryptic statements. The instructions provided with the Central Forge Flaring Kit state:

> The amount the tubing extends above the Die Block (13) will be the finished diameter of the flare.

I confess that I'm not able to parse that statement. Hopefully, you readers will do better than I did.

The best way to get the correct amount of flaring is to understand the requirements and make a test flare or two. We'll be using refrigeration tubing, which comes under the SAE standard J513 (in the United States). This standard sets the outer diameter of the flare mouth for different-sized tubing:

Tubing OD—Width of Mouth

- ¼" = 0.36"
- ⅜" = 0.55"
- ⅝" = 0.77"
- ⁵⁄₁₆" = 0.42"
- ½" = 0.66"
- ¾" = 0.95"

We'll be using ⅜" tubing exclusively for this project. Therefore, we want a tube that looks like the one shown in Figure 9-15 when flared.

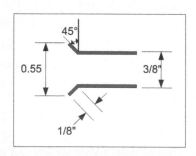

This requires ⅛" of material bent at 45°. The cone will take care of the angle for us; then all we need to do is determine how much tube to leave above the edge of the die for your tool. If your tool has comprehensible instructions, you can just follow them. If not, the following procedure will help you determine the height needed.

FIGURE 9-15: ⅜" Flared tube

Clamp the clean, square, reamed end of the tube flush in the die (no tube extruding past the edge). Tighten the wing nuts to hold the tube firmly in place. Turn the cone assembly sideways and slip the opening across the die and rotate so that the cone is facing the die (on the side with the tube opening). Position the cone so that it is over the end of the tube, and begin turning the cone's lead screw so that the cone enters the tube mouth and the ends of the cone assembly grip the underside of the die.

The cone assembly will twist so that the edges of the assembly are against the edges of the die. This is a good thing and helps center the cone against the die. Once the cone enters the tube, continue turning the lead screw until you meet enough resistance that you can't continue. If you're really strong, don't overdo it. There will be a point where the resistance increases significantly, and this is where you want to stop.

Back off the lead screw and remove the cone assembly. Unscrew the wing nuts and remove the tube. Measure across the mouth of the tube to see if it is 0.55″. This is easiest to do with a caliper, but if you use a ruler, $^{35}/_{64}$″ is close enough for the measurement (a useful trick is to use two drill bits, ½″ and ¾₄″ to get a reference width).

If the width of the mouth is less than that, you need to increase the amount of the tube that protrudes beyond the edge of the die. The length of the bent part of the tube should be ⅛″. Measure that bent part from the outside of the tube at the bottom of where the curve starts. If your flare mouth was undersized, this should be, as well. However much this is short of ⅛″ is how much you should extend the tube beyond the die.

The other end of the tube is easier, since we don't need to unbend it. Once again, place the flare nut onto the tube and flare the end.

Cut the Jets

Cutting the jets is the same process we undertook in Chapter 4, "The Flambeau," and Chapter 5, "The Portable Fire Pit," and you can get extensive details on how to do it in both of those chapters. The refrigeration tubing we're using in this project is softer and easier to work than the hard copper tube in those projects. A few strokes with the edge of a triangular or square file and an easy tap with a brad or tack (you can even do it with a push pin if you're careful) will puncture the jet hole.

We want to orient the jets so that they aim toward the inside rather than straight up. The goal is to surround the vent so that, no matter which way the wind is blowing from, flame from the pilot will engage the propane as it exits. (See Figure 9-16.)

The last step is to add wind protection. Wind is the enemy of consistent ignition. We've already taken one step to address it with the circular pilot. The next step is to

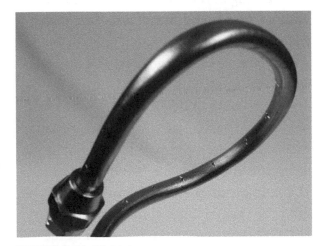

FIGURE 9-16: Pilot jets

wrap the pilot ring with a copper scouring pad. This gives the jets a surprising amount of resistance to wind. Unroll a copper scouring pad (S1) and fold it around the ring so that all the jets have some of the pad above them. Twist the safety wire (S7) around the pad and pilot to hold it in place, as shown in Figure 9-17.

FIGURE 9-17: Scouring pad wind protection

3. Assemble the Base

The 12″ ring at the bottom of the accumulator isn't sufficient to stabilize the boosh in heavy winds or on uneven ground. We're going to reuse the plywood base approach from Chapter 4.

I've recommended a 4′ × 4′ (1.2 m²) plywood sheet as the base. This is larger than most situations require, but it's generally easy to acquire and will definitely secure the effect. To locate the beam clamps, which will hold the base of the accumulator, draw a 6″ × 6″ (152 mm²) square in the center of the sheet. At each of the corners of the square,

drill a ¼″ hole. (See Figure 9-18.) This is where the beam clamps will mount. Screw a ¼″ × ¾″ bolt (Z3) through the hole into each of the beam clamps (Z2) to secure them.

4. Build the Control System

The control box only serves one purpose: to provide 12 V DC on-demand to the main effect solenoid. That's a pretty low bar to meet, but we're going to use three separate switches to accomplish it!

While switching the solenoid on and off (open and closed) is the ultimate function, we have some other needs to meet, as well. The first is to provide an emergency shut-off mechanism. This is a switch that is easy to access, cuts all power to the effect, and requires a positive action to reset. There is a long tradition, in machine tools and processes, to have a "mushroom"-style emergency stop switch to serve this purpose. (See Figure 9-19.)

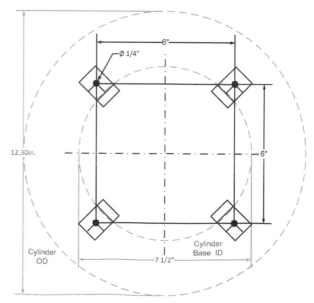

FIGURE 9-18: Bracket hole pattern

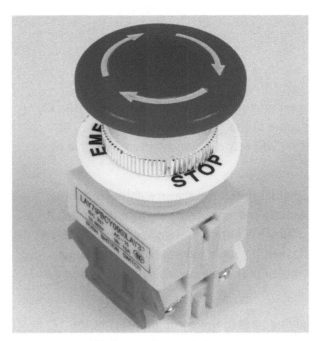

FIGURE 9-19: Mushroom stop switch

There are good design reasons this tradition persists. The red mushroom emergency stop button is visually obvious. It responds to a push, slap, or other simple action. Most switches of this type require the head, once engaged, to be turned before it can be put back in operation. A mushroom stop switch is a great choice for our emergency kill switch.

The second need we have is to disable the system when we walk away. According to NFPA 160 this must be a "removable activator, keyswitch, or coded arming system." By far, the easiest of these to build into our system is the keyswitch. This is a simple electrical switch with a removable key. The switch can only be turned on when the key is inserted. Some key switches will allow you to remove the key when the switch is on. Be careful with these, because they can make it very easy to inadvertently leave the system activated.

Of course, we also need a switch to fire the effect. This should be a momentary contact, normally open switch. Momentary contact means that it is only active when you are pressing it. When you let go, the switch closes. Normally open means that when you aren't pressing it, the switch is open (no power passes through). Switches are sold in normally open and normally closed versions. Some switches provide both capabilities with different contacts. We will only use normally open switches in this project.

The last aspect of the control box is the ability to connect and disconnect it to the power source (12V battery) and the main effect solenoid. It's always worth designing projects to set up and break down easily; hard-wiring components leads to frustration in regular operation and maintenance.

Construct the Switch Box

The choice of what case you use for your switch box is up to you. While you can buy special purpose-built boxes for electronic projects or even fold your own from sheet metal with a box brake, I recommend using a square cookie tin (sometimes sold as candy, cake, or nut tins). You can buy these new online or in grocery or kitchen supply stores. My favorite (and the cheapest) way to acquire these is at secondhand stores. There are always lots of these available in different sizes and at very low cost. The sides of square tins are easier to cut holes in than round tins. The metal that these tins are made of is lightweight and easy to punch or drill through.

Soldering

Soldering with a pencil-tip soldering iron is a powerful skill that you can learn very quickly. This project requires very little soldering, since some of the connections are likely to be screw terminals. But learning to do basic soldering will come in handy when we move on to our Arduino control shield.

For our purposes, any soldering pencil will do. Adjustable temperature soldering pencils, are fantastic, but not a necessity. Getting exactly the right temperature is not a critical factor for most hobby soldering. A low-cost 30–40W pencil will do the job, but a 50W or higher one will get the joint to temperature faster. If you're having to hold the pencil on the joint for a long time and the solder isn't melting, there's probably some other problem going on.

The most important thing about soldering is applying heat to the components, not the solder directly. The soldering iron heats the components being joined; they will melt the solder. The heated metal of the components will wick the solder in, creating a good connection. When the solder doesn't get heated by the components enough to fully melt and wick, a *cold solder* joint is formed. This is visible as dull or rough-looking solder rather than the shiny metal of a good joint.

Always solder with a clean tip. Keep a damp sponge near the iron, and brush the tip across it before soldering. The tip should look like clean metal, not a black or rough surface. With a budget soldering pencil, you can use sandpaper to clean the tip (when it's not hot!). Higher-end soldering rigs are worth cleaning with tip cleaner (a little tub of flux that you rub the tip in).

Soldering should take seconds, not minutes. If you're waiting more than 60 seconds (depending on the size of the joint), heat isn't transferring properly. A clean tip is the first thing to check for. The other essential factor for good heat transfer is to coat the tip with a thin layer of melted solder. This is known as *tinning* the tip.

When you're ready to solder, having wiped the tip clean, hold the solder against the tip so that it melts and coats the end. You don't want a big goopy blob hanging off; if that happens, wipe it off on your sponge. You want the tip coated in shiny solder. This is the part of the tip you'll hold at the base of the components. You'll know that the pencil is heated up when it coats quickly and easily; if it takes any time to melt the solder, the tip isn't fully heated.

With the clean, hot, tinned tip at the base of your joint, hold the end of the solder across the components being soldered. In a couple seconds, you should see the wires or terminals melt the solder and wick it in. Once you see this, you can stop heating the joint. If the solder doesn't melt, don't keep heating so long that you damage the components. Stop and clean the tip and try again.

Pro tip: At some point, you will forget this next tip. When you do, you'll get your membership in the vast community of those of us who have stared in frustration (and perhaps uttered expletives) as we realize that we soldered the joint before putting the heat shrink tubing on. Always remember to put the heat shrink on before you solder. When you do forget, just remember you're in good company!

Plastic containers will work, as well. I've used old lunch boxes, VHS tape cases, and Tupperware as project cases in the past with great success. The only real requirements are the ability to make clean holes in the material, the ability to close the case up, and a finished case that is robust enough to hold up to working conditions.

The shape should be something that you can comfortably hold in one hand while operating with the other. The case needs to be deep and wide enough that the internal parts of the switches fit without rubbing or touching the sides or bottom. Don't worry if the tin or case has a print of holly leaves and berries or is plaid or has flowers on it; we'll paint it before we're done (unless you like the pattern or color).

Pro tip: Over the years I have developed the strong opinion that, when building project cases, if the case has two parts (box and lid, two matching U shapes, etc.) you should only attach components to one of them. The value of doing this is that disassembly and

troubleshooting are vastly easier if you can take off the lid, or the box, and not have to fiddle with wires or cables that are holding it to the other part of the case. If there is no way around this, do it with as few parts as possible, and always include an easily releasable connection on any wires that go between the two case parts.

The design I'm presenting turns the tin upside down. It has all the switches mounted to the bottom of the tin. The incoming and outgoing connectors go through the sides of the tin. You can remove the lid (now the bottom) and gain access to all parts, even allowing troubleshooting with a multimeter while power is connected. Feel free to build it in a different manner, but always keep in mind how you will be to get access for troubleshooting if you need to (and you always will).

Drill the Holes

While your component and case choices may be different, I'm going to offer a design based on a 6″ × 8″ × 2″ tin (C4). The size is arbitrary; any reasonable size that will hold the switches will work. That's just the only size I could find at the secondhand store. Two inches is probably the shallowest case that will reasonably work. I'm also going to use some of my favorite switch components. Keep in mind that there are hundreds of functionally equivalent alternatives; just pay attention to the dimensions and you can swap out any momentary contact, keyswitch, or emergency stop switch for the ones I used.

We will need to drill the following holes, as shown in Figure 9-20:

- Incoming power: ⅜″
- Main effect: ⅜″
- (C2) Keyswitch: ¾″ (modify to meet your switch if needed)
- (C3) Mushroom switch: ⅞″ (modify to meet your switch if needed)
- (C1) Effect switch: 1″ (modify to meet your switch if needed)

Drilling thin metal like this can be frustrating. Ragged edges and imperfect holes are not uncommon results. My recommendation is to use a step drill with a scrap piece of 2 × 4 backing the metal to be drilled. Ideally, you will clamp the wood to the tin, but if that isn't feasible, use your heavy gloves and hold the tin tightly against the wood while drilling. Switch the side you're drilling from (inside to outside and vice versa) with each step of the drill to get a cleaner hole.

Use the center punch to mark the drillable locations per the diagram above. The wider step bit will go all the way to the 1″

hole we need to drill. If you have access to hole saws or spade bits of the correct size, you can use them. With all types of bits, you *must drill slowly*. You do not want the drill to bite hard, grab the metal, and turn the tin into a spinning menace.

Paint the Tin (Optional)

It is by no means a requirement, but if you want to paint the tin, this is the time to do it. Put the lid on the tin and paint it and the sides. You can paint the bottom (where the switches are) if you want, but it's likely to get

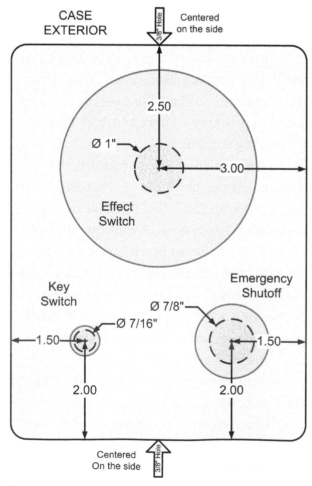

FIGURE 9-20: Control box drill layout

scratched, so you'll want a couple coats of good hard enamel.

I like an industrial look to my controllers, so I always wrap the sides in black and yellow reflective safety tape. If the tape is wider than the tin walls, you may have to use a razor knife or something sharp to cut it away from where the lid will go on. You can paint this pattern by using strips of wide masking tape at a 45° angle with yellow on top of black. Space them as far apart as the tape is wide. (See Figure 9-21.)

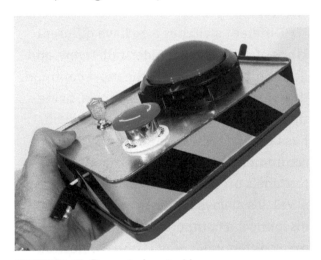

FIGURE 9-21: **Decorated control box**

Mount the Components

The three switches are mounted by unscrewing the retaining nut, pushing the body through the hole, and then screwing the retaining nut on from the other side and tightening. Different switches come apart in different ways.

Most contemporary mushroom-style emergency stop switches come apart for mounting by unscrewing the red mushroom head, then unscrewing a collar that threads around the body. The body pushes through from underneath (unlike the other switches); then the collar and head are put back on and tightened. Most other switches are mounted by pushing their bodies through from the top and rethreading the retaining nut on the inside. (See Figure 9-22.)

FIGURE 9-22: **Mounting switches**

The ragged sharp edges of the two holes on either end of the box, used for power input and main effect control output, would be dangerous to run wires through unless we do something to buffer them. Luckily, rubber grommets (C6) are made specifically for this purpose. They're shaped like a torus (donut) with a slit around the edge. The grommet bends to fit into the hole and the metal of the wall fits into the slit. The rubber passageway protects wires from the sharp or jagged metal. To mount these, fold one edge of the torus inward so the grommet looks like a C instead of an O. Push one side of the material into the hole and let the grommet unfold. (See Figure 9-23.)

FIGURE 9-23: Mounting grommets

Wire the Components

It's likely, if you used components similar to mine, that the momentary contact switch and the emergency stop switch have both normally open (NO) and normally closed (NC) connections. The difference is that an NO switch does not pass electricity until the switch is closed. An NC switch does the opposite; it passes electricity until the switch is closed. We will use only the NO mode of switches in this circuit. Switches should be labeled with NO or NC. Sometimes there are three contacts: a common (COM) contact and one NO and one NC contact. Other times there are independent NO and NC sets of contacts.

You've probably noticed by now that I'm using a *huge* momentary contact switch (C1). This switch is from arcade systems and is irresistible for people. It even lights up! Pushing the "big red button" to fire a flame effect adds an extra-special boost to the experience. These switches are sold online at electronics sites like Adafruit and SparkFun as well as Amazon.

Lighting the switch is relatively easy; there are two contacts that will light a resistor/LED combination. These are usually designed for 5V, but some online sources state that they will work up to 12V. Because the LED is a diode, it will only light if the correct polarity is applied; in other words, one terminal is positive and the other is negative. It can be difficult for many lighted switches to determine which is which without testing.

Different switches also have different kinds of connectors; solder-tail, screw, and flat-crimp terminals are the most common. If your switch has screw terminals, using them as intended is ideal. If your switch has tab-type crimp connectors, I highly recommend using a wire crimper to put terminals on the ends of the connecting wires. It's always better to be able to easily disconnect a component. However, there are many different sizes of crimp connectors and I'm very hesitant to add yet another tool for readers to buy. So I'm going to treat the crimp terminals on the switches in this project as solderable (which they are) and solder the wires to them.

There are many choices for hook-up wire to use between components. Arguments can be made for solid core and stranded wire. While it's a bit of overkill for the current we'll be dealing with (generally less than 2 amps for a ¾" 12V DC solenoid) I like to use 16-gauge stranded wire for this kind of work. I'm going to encourage tinning the ends of the wire. This involves melting solder into the strands so that they form a solid unit. This overcomes the frustrations that can come with screwing down or connecting stranded wire. Stranded wire also withstands movement over time better than solid core. While we hopefully won't have much movement in the hookups between the components in this box, it is intended to be a rugged kit that will serve you well for years.

The wiring for this controller is very simple. Cut one of the two-pin trailer hookup connecters (C8) in the middle of the leads. Strip ½″ of insulation off each lead.

Bring the leads in, through the grommets, from the two-pin connectors. On the inside of the tin, tie the leads in an overhand knot (or better yet, a square knot). This serves as a strain relief so that tension put on the connectors doesn't pull the wires inside the tin. (See Figure 9-24.)

The wiring is very simple for this project. (See Figure 9-25.)

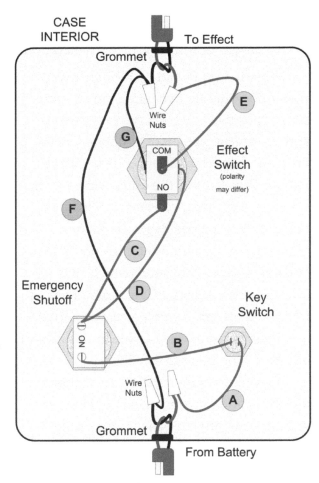

FIGURE 9-25: **Wiring diagram**

FIGURE 9-24: **The trailer hookups ready in the control box**

The positive input will be routed through our three switches in series:

A. Positive input to key switch

B. Key switch to emergency shutoff (NO)

C. Emergency shutoff (NO) to effect switch NO terminal

D. Emergency shutoff (NO) to positive LED input

E. Effect switch COM terminal to positive output

The negative input will route from:

F. Negative input to negative output

G. Negative output to negative LED input

Because we are cutting a single two-pin trailer hookup into two parts, the relationship of positive and negative to the exposed and concealed pins on the hookup will be different on each half. One side will have the red pin exposed and the other will have the black pin exposed. I consider it a best practice to shield the positive (red) pin on outputs. This derives from the common practice of using a negative ground (which isn't relevant to this project), so if a bare pin has to touch something, it won't cause a short. We aren't using a grounded frame in this project, but it's a good habit to establish.

We will use wire nuts (C7) to connect the incoming and outgoing wires to the internal wiring. Wire nuts are easy to use. Strip ½" (1 cm) off the end of the wire. We don't want to tin the tips of wires that are used in wire nuts—the nut will twist the strands together for the best connection. Twist the stripped ends of the two or more wires to be connected then insert them into the nut. Twist the nut around them until you feel it bite and take resistance. Twist it firmly, but not so hard that it breaks loose.

If there are screw connections available, use those for connecting wires. Otherwise we'll solder the connections. Strip ½" (1 cm) of insulation off the wire. Twist the strands tightly together. Clean your hot soldering iron by wiping it on a damp sponge and tin the end of the iron by holding the solder against it until it melts and leaves a shine on the tip.

Tin the tip of the wire by holding the tinned tip of the soldering iron at the base of the stripped section and the solder on the other side, closer to the end. The solder should melt and flow into the wires. You only need a small amount of solder in the strands to tin them; too much and it becomes bulky.

The contacts on the switches may have holes in the lugs or tabs. Slip the tinned end of the wire through the hole (this is why you don't want it to be bulky!) and bend it back against itself, twisting if possible (I'll describe this as a *mechanical connection*). Be sure that it is positioned so that it won't touch anything else (especially the other contact or wire!) once soldered.

On occasion, no hole exists, or it is otherwise impossible to pass the wire through the lug or tab. If you can't wrap the wire around the terminal, you can still solder it by resting it against the tab or lug. This can be tricky without what is called a *third hand*. This is an alligator clip on the end of an armature that will hold the wire steady for you while you solder. Once there is a good solder joint between the two parts, the solder will hold the wire in place.

With the wire mechanically connected, hold the tinned tip of the soldering iron on one side of the joint, ideally touching both the wire and the terminal tab. Hold the solder on the opposite side. Once the solder melts and flows, we'll have established a reliable *electrical connection*.

We all know people who will want to push the buttons so hard that the contacts might touch the other side of the control box. To avoid this, put some duct tape or vinyl down on the inside of the lid. It only needs to cover the area under the switches, but remember that the lid can go on either way, so go ahead and cover most of it. If you can find one, a deeper box would avoid this altogether.

Solder or screw all connections as described and hopefully your system will look something like the one shown in Figure 9-26 when complete.

Wire the Connectors

The connectors I'm specifying are generally used for trailers. You can buy them online, at auto parts stores, and at some discount import tool stores. These connectors are

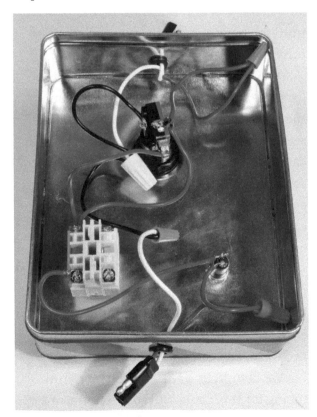

FIGURE 9-26: Completed wiring

relatively cheap, handle a lot of current, and are designed to deal with weather and hard use. They come in many pin configurations, but the two-pin model is especially useful for power transmission.

Wire the Solenoid

The solenoid is easy—simply strip the ends and use wire nuts to attach the two-pin connector's wires to the solenoid wires. As we noted earlier, the solenoid's inputs are not polarized. It will work regardless of how the wires are connected. I recommend putting a zip tie around the body of the solenoid and wires to act as a strain relief. Wires get tugged, so you always want to make sure that, when they do, they don't tug on the electrical joint. (See Figure 9-27.)

Wire the Power Source Cable

The other connection is to the power source. This needs to be a 12V DC source that can handle repeated brief 1.5 amp loads. A car battery would be heavy to carry around (and overkill). A motorcycle battery would work, but you'd need a mechanism to charge it.

FIGURE 9-27: Strain relief on the solenoid

Various scooter batteries are also a workable choice, but they too need a charging mechanism.

Over the years I have happily settled on a solution that I strongly recommend. Portable jumper boxes, or jump-starters, provide a stout battery and a built-in charging mechanism. Many also include an air compressor and light. These boxes will easily provide a weekend's worth of booshing, even with multiple solenoids. They also have the advantage of being incredibly useful when you aren't booshing and need a spare power source for camping (or even their intended purpose of jump-starting a dead vehicle). They range in price; I've bought a number of them over the years. I really like having the built-in air compressor, but it's not required for the use we'll be putting it to. Check the sales at your discount import tool store and online sources for really good deals on these.

We need to fabricate a cable from the power source to the control box. On the control box side, we will use one of the two-pin power connectors. On the source side we need whatever will connect safely to the source. Jumper boxes should all come with a 12V cigarette lighter socket. Our cable will put a cigarette lighter plug on the source end. If you've gone another route with your power source, you may need spring clamps or some other connection.

You can purchase a replacement cigarette lighter plug online at places like Amazon. Or you can look around for a broken or spare car charger with a plug. Whichever way you acquire one, make sure the leads coming

out of it are at least 16-gauge wire. Many car chargers have very lightweight cables and are not appropriate for heavier current.

Take the replacement cigarette lighter plug (C9) and strip the last inch (2.5 cm) of wire on each lead. If the wire that the factory attached to the cigarette plug isn't long enough, at least 4′ (1.2 m), you can splice in wire to make it longer. Any 16-gauge or heavier wire will do. For cost effectiveness, I recommend using lamp wire, which is just standard AC two-wire cable. You can usually purchase extension cords for less than the wire itself. The minimum useful length for this cable is 4′ (1.2 m) but anything between 4′ and 8′ (1.2–2.4 m) will work well. If necessary, you could make this considerably longer, but you don't want to be too far away from the power source in case you need to turn it off quickly.

You'll need to determine the polarity (positive and negative wires) coming out of the cigarette plug so that you connect correctly to the three-pin trailer connecter. You can do this with a multimeter with the cigarette plug plugged into the jump box, or you can use an ohm meter between the outer contacts of the cigarette plug (negative) and one of the wires.

Take the wire attached to the cigarette plug, or the added lamp cord (C5), and peel the pair of wires apart for about 6″ (15 cm) on each end. Strip 1″ (2.5 cm) of insulation off each wire on both ends of the cord.

We definitely want to use heat shrink tubing for this cable. Cut four lengths of 1½″ (4 cm) heat shrink large enough to loosely fit over the cable. Slide one of these segments as far as it will go onto each of the ends of the power cord. Cut two 3″ (7.5 cm) sections of larger diameter heat shrink tubing and slide them over both pairs of wires onto the unseparated cable. (See Figure 9-28.)

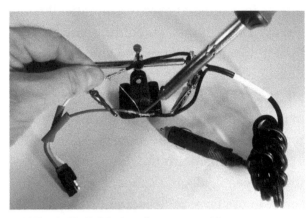

FIGURE 9-28: **Soldering the power cable**

A lineman's splice consists of a simple set of twists of the two wires to be joined. Let's start with the negative wire from the cigarette lighter plug connecting to the negative wire from the lamp cord.

1. Loosely twist the strands of each wire so that they aren't splayed out.

2. Cross the two wires like an X with the crossing point ¾″ of the way toward the insulation.

3. Twist the two wires around each other (not just one around the other one) 2 times.

4. Wrap the remaining length of each wire around the other wire.

(See Figure 9-29.)

Once the splice is complete, solder the length of the splice. Slide the heat shrink

down over the splice and use a lighter or heat gun to shrink it. Be sure to cover all of the exposed wire. If the heat shrink is too short, use electrical tape to cover any remaining wire.

Perform the splice, solder, and heat shrink steps for the other wires. Be careful to note the polarity on the power cord to make sure you connect the two-pin trailer plug for the control box correctly. Once you have all the splices completed and insulated, slide the larger section of heat shrink down over the pair of splices on each and shrink it in place.

Wire the Effect Cable

Follow the same process to fabricate the effect cable as with the power supply cable. The only differences are that there is a two-pin trailer connector at each end, the polarity doesn't really matter, and we want a longer run of cable. This cable should be at least 12′ (3.75 m).

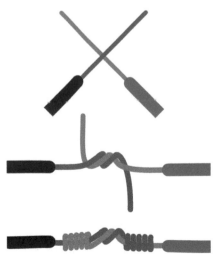

FIGURE 9-29: **Lineman's splice**

5. Final Assembly for Testing

We're in the home stretch. Now all we have to do is put the pieces we've fabricated together. But before you do . . .

Paint the Effect

In truth, this is an optional step. However, there are a number of surprising benefits of painting the iron and steel in your boosh. I cannot put paint on copper or brass; it is an affront against a truly lovely metal, but the black iron pipe and the steel of the accumulator cylinder are prime targets for sprucing up.

Black iron pipe rusts. The black finish on the pipe itself retards some rusting, but the threads and the fittings will turn positively orange shockingly quickly. Painting iron pipe is the traditional way of keeping rust at bay (ask any sailor). Showing up for a permitting inspection with rusty kit is not a great way to kick things off. As a bonus, you can store the equipment outside without having to wire-brush it every time you want to use it.

Painting the cylinder is a little more complicated. You may have to use some solvent and a good deal of elbow grease to remove all the labels. I confess to using fairly strong solvents to remove all traces of labels, because I have invested in the appropriate safety gear. Always research the protection requirements for any solvent you consider. I am very happy with the investment I made in an organic vapor cartridge respirator. It, along with double nitrile gloves and eye protection, greatly extends the range of solvents I am comfortable working with. Trading elbow grease with a less-toxic solvent is also always a good deal.

The other advantage of painting is less direct. My experience with having my flame effects inspected at events dramatically changed when I started painting my gear. I paint my accumulators red, and my iron pipe and fittings blue for plumbing and black for exhaust. The reaction from fire department inspectors was startling: they took one look at the clean, professional appearance the paint provided and relaxed to a noticeable degree. Inspections that used to get to the level of "How many wraps of Teflon tape does that joint have?" became "How often do you take this effect out?" I can't guarantee that everyone will have this experience, but I have to say that even other propane enthusiasts have stopped and complimented my kit when they see the painted plumbing. It just flat out makes you look more professional.

Painting before assembly is a recipe for having to do touchups. Assembly causes scratches. My recommendation is to assemble the boosh and then spend time with masking tape and newspaper to protect areas that you don't want to get paint on. Cover up the brass and cylinder and paint the iron pipe and fittings. Cover everything else up and paint the cylinder. The color scheme is your choice, but you're welcome to the red and blue or red and black motifs I use; I certainly can't claim them as my own. (See Figure 9-30.)

FIGURE 9-30: **Painting the effect**

Assemble the Pilot Mounts

We will use the same kind of conduit hanger we used in the flambeau stand to mount the pilot to the main effect pipe. However, since we are using ⅜″ tubing for the pilot, there are no matching conduit hangers of that size. Instead we'll mate a ½″ hose clamp to the ¾″ conduit hanger to attach the tube to the pipe.

We will make two of these. To assemble the first one, unscrew the hose clamp (S5) and stretch it out flat. Drill a ¹³⁄₆₄″ hole in the clamp approximately in the middle of the metal strip. You can use the tail of a file or a small screwdriver blade to ream it out enough to accept the ¼″ bolt.

Twist the ¼″ × ½″ bolt (S2) through the hole and through the back of the ¾″ conduit hanger (S6). Thread a ¼″ lock washer and nut (S3, S4) onto the bolt and tighten. The result should look like Figure 9-31.

Fabricate one more mount in the same manner. We'll leave the hose clamp open until we put it around the pilot tube.

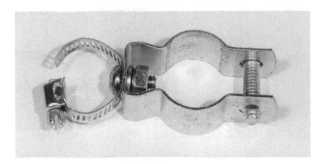

FIGURE 9-31: **Assembled pilot mount**

Attach the Pilot Tube to the Main Effect Pipe

Unscrew the tightening bolts from the ¾″ conduit hangers and slide the hangers over the ¾″ × 6′ black iron main effect pipe (P1). Put the bolt back in and tighten so that one hanger is about 12″ (30 cm) from the top of pipe and the other is about 18″ (45 cm) from the bottom.

Set the pilot tube against the mounts so that the ring is more or less at the end of the pipe. Wrap one of the open hose clamps around the tube and insert it into the tightening mechanism. Turn the tightening screw until the hose clamp binds tightly around the tube. Do the same thing for the other clamp. (See Figure 9-32.)

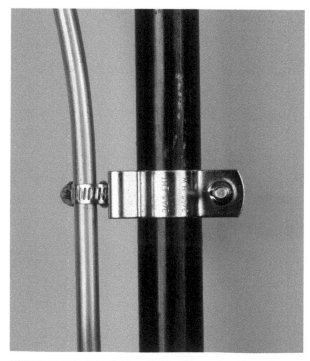

FIGURE 9-32: **Attach the pilot to the main effect pipe.**

Testing

This section describes testing everything except an actual live ignition of the effect. This is a test you can perform even if you are in an area where it is not safe or legal to ignite the effect.

System Assembly

If you didn't store the boosh with the stand attached, attach the accumulator with the plumbing tree to the brackets on the stand before you begin the system assembly.

Attach the Pilot and Main Pipe

Thread the main effect pipe into the solenoid. Be careful to not disturb the pilot vent tube attached to it more than necessary. (See Figure 9-33.)

Once the main pipe is securely threaded into the solenoid, adjust the pilot tube so that it is on the side of the tube with the low-pressure regulator. Be sure the pilot tube is parallel to the main pipe and the ring on top surrounds the main vent. The pilot ring should be about 1″ higher than the mouth of the main vent.

Tighten the conduit hanger on the main pipe and hose clamp on the pilot tube, if needed. Thread the ⅜″ FFL fitting on the end of the low-pressure regulator onto the ⅜″ MFL fitting at the bottom of the pilot tube's cutoff valve. Tighten this appropriately. Check to make sure the needle valve and the pilot cutoff valve are closed.

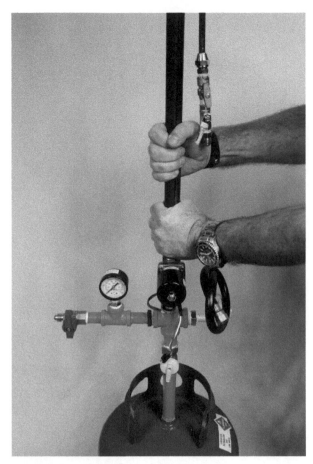

FIGURE 9-33: **Attach the main effect pipe.**

Attach the Supply Hose

Check to make sure the accumulator shutoff and main inlet valves are closed. Attach the ⅜″ FFL fitting on the main supply hose onto the ⅜″ MFL fitting on the end of the main gas inlet.

Connect the Control System

Turn the key switch to the off position and remove the key. Engage the emergency shutoff switch. Connect the power source cable to the power inlet connector on the control box. Connect the effect cable to the solenoid and the control box. Insert the cigarette plug into the outlet on the jumper box (or however you constructed a connection to the battery).

Connect to the Cylinder

Thread the cylinder POL fitting into the cylinder outlet. Verify that the source cutoff valve is closed. If you haven't added the additional hose (E1) to the high-pressure source to have at least 15′ (4.5 m) between the fuel and effect, do so now.

Connect the ⅜″ FFL fitting on the end of the extended supply hose to the ⅜″ MFL fitting on the boosh's gas supply inlet. Verify that the gas inlet supply cutoff valve is closed.

Leak-Testing

Starting with all valves closed and the solenoid disconnected from the control box, open the cylinder valve fully and then back a quarter turn. Adjust the high-pressure regulator to 10 psi. Open the source ball valve, effect feed valve, accumulator valve, and pilot cutoff valve (leaving the pilot needle valve closed). Verify that the effect gauge matches the regulator gauge reading.

Leak-test all fittings and hoses. If leaks are detected, close all valves, tighten, and restart the test procedure.

An excellent additional test is to close the source cutoff valve and see if, over the course of 10–15 minutes, the boosh gauge drops. If it does, you've got a hidden leak. Aggressive leak-testing is in order. You can try this at various pressures.

Pilot System Testing

Prepare a torch or fireplace lighter to light the pilot. You may need a stool or step ladder to reach the pilot. *Do not*, under any circumstances, climb on the boosh or rely on it for balance when attempting to light the pilot.

Open the gas inlet valve if it is closed. Open the needle valve all the way, light your torch or lighter, and position yourself to reach the jets. Open the pilot cutoff valve and light the jets. Adjust the needle valve to the lowest level of flame that will stay lit with whatever wind is present. (See Figure 9-34.)

Safety Note

The tests following the pilot system test will discharge unignited propane. Do not perform these tests indoors or in any space without excellent ventilation. Do not perform these tests around any source of ignition such as flames, pilot lights, cigarettes or smokers of any kind, hot electrical circuits, or other heat sources. Remember, propane is heavier than air and will sink to ground level. Sufficient ventilation (or wind) to disperse the propane is essential.

FIGURE 9-34: The pilot light ignited

If the jets don't light, step down and shut off all valves. If the system has pressure and the effect input valve is open, then you either have a bad low-pressure regulator or the line is blocked. If needed, you can connect the LP regulator hose directly to the pilot vent tube with a male-male ⅜" flare adapter and test the regulator directly. This is only useful for testing. The needle valve and pilot cutoff are essential for regular use of the boosh.

Troubleshooting the Control System

You may need to test the circuit with a multimeter if there are problems. Test the following in sequence to troubleshoot:

1. In voltmeter mode, connect the multimeter leads to the solenoid end of the effect cable and press the main effect fire switch.

   a. If you read 12V DC, then check the connector that's soldered onto the solenoid leads. Worst case (and very rarely), the solenoid is bad.

   b. If you don't read 12V DC, then go to the next step.

2. Disconnect the effect cable from the control box and test the contacts on the output of the control box when the main effect fire switch is pressed.

   a. If you read 12V DC, the effect cable is bad and should be torn down and remade.

   b. If you don't read 12V DC, then go to the next step.

3. Unscrew the wire nut on the negative lead where it connects to the effect cable connector, and use the exposed wires as the connection for the negative probe from the multimeter.

4. With the positive probe from the multimeter, test the output lead on the main effect fire button when you press the switch.

   a. If you read 12V DC, then the connection from the main effect fire switch to the output connector is bad; check the wire nut.

   b. If you don't read 12V DC, test the input lead on the main effect fire switch.

      i. If you read 12V DC, then the main effect fire switch is bad.

      ii. If you don't read 12V DC, go to the next step.

5. With the positive probe from the multimeter, test the output terminal on the emergency shutoff switch.

   a. If you read 12V DC, then the wire between the emergency shutoff switch and the main effect fire switch is bad; reconnect and/or re-solder.

   b. If you don't read 12V DC, then test the input lead on the emergency shutoff switch.

      i. If you read 12V DC, then the switch is bad (or possibly accidentally wired to the NC terminals).

      ii. If you don't read 12V DC, go to the next step.

6. With the positive probe from the multimeter, test the output terminal on the key switch.

   a. If you read 12V DC, then the wire between the key switch and the emergency shut-off switch is bad; reconnect and/or re-solder.

   b. If you don't read 12V DC, then test the input lead on the key switch.

      i. If you read 12V DC, then the switch is bad.

      ii. If you don't read 12V DC, go to the next step.

7. Disconnect the power cable from the control box and test for 12V DC with both multi-meter probes at the end of the power cable (where it would connect to the control box).

   a. If you read 12V DC, then

      i. The connector on the control box's power input is bad (check the wire nut).

      ii. The negative wire leading from the power connector to the effect connector is bad (check the wire nuts).

   b. If you don't read 12V DC, then one of these is the culprit:

      i. The jumper box isn't turned on (turn it on).

      ii. The jumper box isn't charged up (charge it up).

      iii. The cable is bad (tear down and rebuild).

      iv. The cigarette plug is bad. (Many cigarette plugs have a fuse, so check it to see if it is blown.)

There are other possibilities for problems in the control system, but this gives you a pretty good idea of how to methodically troubleshoot a problem. Always break your testing down into a sequence of actions that isolates segments for testing.

When pilot testing is complete, close the pilot cutoff valve and verify that the pilot extinguishes itself after a minute or so. If the pilot does not extinguish itself, there is a problem with the pilot cutoff valve.

Control System Testing

Verify that the pilot is *not* lit before performing this test. If the pilot has been recently lit, wait three to five minutes for the pilot to completely cool before proceeding. Verify that the pilot cutoff valve is in the closed position.

Insert the key and turn the key switch to the on position. Twist the emergency shutoff switch so that it disengages. Verify that the area is clear and free from ignition sources.

The system should still be pressurized after performing the pilot test. Verify that the gauge on the effect shows pressure. Close the cylinder valve all the way to make sure that no more pressure enters the system.

Press the main effect fire switch. If the solenoid engages, you will hear a loud startling *bang*! as the solenoid opens and the gas ejects under pressure. (If the solenoid doesn't engage, see the "Troubleshoting the Control System" sidebar.) Engage and hold the main effect switch to open the solenoid and release all the pressure in the accumulator. You will be able to hear when all the pressure has been released.

When finished with the control system test, engage the emergency cutoff switch, turn the key switch to off, and remove the key (put it somewhere safe!).

Shut Down the System

Verify that the cylinder valve is completely closed. Close the source cutoff valve and gas inlet valve. Verify that the pilot cutoff valve is closed.

Disconnect the power cable from the power source and the control box. Disconnect the effect cable from the solenoid and control box.

Operation

Igniting the boosh is not something you can do covertly. If you live within city limits or in some other area that prohibits open flame, firing a 6′ × 15′ fire column is a great way to get a visit from law enforcement or your local fire department. *Do not* fire this effect in locations where it is prohibited. The laws, codes, and regulations vary so much in different places that I cannot provide advice other than to contact your local fire department and ask them where testing a flame effect might be permitted.

Your odds of a good reception from the fire department increase dramatically if you have good documentation about your project. Provide a schematic and block diagram, and write up an operating procedure and safety plan. Comb your hair and brush your teeth. Attempt to look like a fine, upstanding citizen who loves mom, safety, and apple pie. Many firefighters share a love of flame. If you come across as reasonable and safety conscious, you may find kindred souls who can be a tremendous help.

The operational directions provided below are for your personal use. They are not sufficient for operating before an audience. See the "Writing a Flame Effect Permit Plan" sidebar for more information about lighting up before an audience. The definition of an audience varies from jurisdiction to jurisdiction. In Texas, it's 50 or more people (Texas Occupations Code, Title 13, Subtitle D, Chapter 2154, Subchapter A, Sec. 2154.253 (b).). You need to research what the requirements are for your location.

Component Checklist

Below is a checklist of what you need to have when taking the boosh somewhere or bringing it back. (You can also see what it looks like in Figure 9-35.)

FIGURE 9-35: The boosh kit

Set Up the Environment

The first thing to do is to make sure that the environment is under control. This means that there is nothing in the area that will catch fire or crash into the effect. It also means that the fuel (propane cylinder) is sufficiently far away from the effect (a minimum of 10′, and more is better) and in no danger of being knocked over. You need to be sure that no one will inadvertently wander into the operating area.

Verify your hoses are not in a traffic area. Make sure your extinguishers are placed in appropriately accessible locations. Make sure no people, property, trees, or anything flammable are in a position to receive excessive heat from the effect. While I've rarely seen anyone actually measure this, NFPA-160 states that the temperature of skin exposed to the effect for 4 seconds will not exceed 111°F (44°C). Combustible materials cannot exceed 117°F (47.2°C). You'll be surprised at how much heat you'll feel from the boosh and how far away from the effect you'll feel it.

You are responsible for the effect. The project described above includes a key lock. Keep control over the key and only insert and arm the effect when preparing to operate. Do not leave the effect armed and unattended.

You also need to have a means of extinguishing anything that catches fire.

ITEM
Propane Cylinder
High-pressure source with extended hose
Control box
Power source
Power cable
Effect cable
Control box key
Effect base
Effect accumulator/plumbing tree
Effect main pipe/pilot
Crescent wrench
Pilot ignition source
Fire extinguisher

Fire Extinguishers

Fire extinguishers are rated for the types of fires they are safe to use on. In the United States, type A is for wood, trash, and paper. Type B is for flammable or combustible liquids and gases (including propane). Type C is for electrical fires. Other countries use different classification systems, so be sure to understand which class is rated for propane in your area.

In general, there are three types of extinguishers that are commonly used around propane (foam is a runner-up, but not as common as the three below):

- Dry chemical
- Pressurized water
- CO_2

Dry chemical extinguishers are by far the most easily accessible and common type of extinguisher. Most carry an ABC rating, making them technically appropriate for all types of fires. They are the least expensive option for extinguishers and come in both rechargeable and disposable models. If you've never had to discharge an ABC dry chemical extinguisher, they sound ideal. The reality is somewhat different.

ABC dry chemical extinguishers are effective, but the most common powder used, a yellow mix of monoammonium phosphate and ammonium sulfate, is an extremely unpleasant thing to get coated with or breathe in. The powder can irritate the respiratory system, eyes, and skin. It's also corrosive and needs to be cleaned off metal and other surfaces (and there is a *lot* of it once you start spraying). While getting coated with the powder in the ABC dry chemical extinguisher is far better than getting burned, it is not something that you would ever want to have happen unless that was a serious threat.

The NFPA 160 *Standard for the Use of Flame Effects Before an Audience* (11.3.2.1) states:

> In all cases, at least two pressurized water, Class 2-A, extinguishers and two Class 10-BC extinguishers shall be provided, in addition to those required by NFPA 10, Standard for Portable Fire Extinguishers, for the building.

I've had fire inspectors reject ABC extinguishers from meeting this criterion (which I now agree with). The reason is because pressurized water is easier and safer to use around people and on clothing, and 10-BC, which is generally either CO_2 or sodium bicarbonate, is far less offensive than an extinguisher with the full ABC rating. Most firefighters love CO_2 extinguishers, because they're quick and easy to use, but they are frustratingly expensive for the average user and have to be fully charged to be effective (which is also expensive).

In a situation where you are not in front of an audience, a garden hose can also be a great resource to have on hand in case of small type A fires.

The mnemonic for using a fire extinguisher safely is PASS. P is for **P**ulling the pin. A is for **A**iming at the base of the fire. The first S is for **S**queeze the lever slowly, and the last S is for **S**weep from side to side.

Firing the Effect

Assemble the effect per the system assembly instructions described in the "Testing" section above.

1. Prep the fuel

   a. Open the main effect valve.

   b. Open the accumulator cutoff valve.

   c. Open the cylinder valve all the way and back it off a quarter turn.

   d. Open the regulator to the desired pressure (start off with 40 psi) using the regulator gauge to measure.

   e. Open the source cutoff valve (you will hear the system charging).

2. Ignite the pilot.

   a. Open the needle valve all the way.

   b. Prepare your pilot ignition source (propane torch, fireplace lighter, etc.) and the means by which you will reach the pilot ring (long arms, footstool, box, etc.).

   c. Open the pilot cutoff valve.

   d. Light the pilot ring with the ignition source.

   e. Reduce the needle valve setting to the smallest flame that the wind won't put out.

3. Arm the effect.

   a. Verify the location is ready, with no people near the effect.

   b. Put the key in the key switch and turn it to on.

   c. Disengage the emergency cutoff switch.

4. Fire the effect.

   a. Push the main effect fire switch.

5. Whoop for joy! (See Figure 9-36.)

FIGURE 9-36: **Boosh!**

Safety Note

The two primary means of rapidly shutting the effect off are the Emergency Shutdown switch and the source cutoff valve. At the first sign of any danger, engage whichever you can as quickly as possible and the other directly after.

Follow by closing the cylinder valve. If, in some extreme circumstance, the cylinder or regulator catches fire, leave it alone, evacuate the area, and immediately call the fire department.

Shut Down the Effect

To shut the effect down in an orderly manner:

1. Cut off the gas supply.

   a. Close the cylinder valve completely.

   b. Vent the accumulator pressure by pressing the main effect switch. This will create one final flame; close the pilot cutoff valve and let the pilot fully extinguish prior to this if you do not want a final flame at this stage.

   c. Close the source cutoff valve.

   d. Close the gas inlet valve.

   e. Close the accumulator shutoff valve.

   f. Close the pilot shutoff valve.

2. Disarm the controls.

   a. Close the emergency cutoff switch.

   b. Turn the key switch to off and remove the key.

   c. Turn off the power supply if it has a switch.

System Disassembly

Allow the boosh to cool down prior to disassembly. Only the top area of the pilot and the upper part of the main vent pipe should be hot, but always handle them with care.

Break Down the Controller

Having turned off the key switch and removed the key (place it somewhere safe) per the instructions above, unplug the power cable and the effect cable from the controller, solenoid, and power source.

Break Down the Gas Supply

Disconnect the supply hose from the boosh's main inlet. Unthread the regulator from the cylinder.

Break Down the Boosh

Disconnect the low-pressure regulator hose from the pilot tube. Unthread the main effect vent pipe from the solenoid, leaving the pilot tube attached.

Writing a Flame Effect Permit Plan

If you find yourself interested in firing your boosh before an audience, you may encounter the need to get the effect permitted. The specific requirements for this vary tremendously from location to location. The requirements can even vary dramatically depending on the specific people involved at a location.

What one jurisdiction, or even one individual at a jurisdiction, considers an absolute requirement, another jurisdiction may not care about. Other than NFPA-160, *Standard for the Use of Flame Effects Before an Audience*, there are no national standards that everyone consistently follows. Different locations, inspectors, or licensed flame effect operators (FEO) are free to add their own requirements.

I will talk about the requirements in Austin, Texas, since that's the location where I do most of my FEO work. You must do the research to understand who the authority having jurisdiction (AHJ) is in the locale you're considering. Once you know who the AHJ is, you then have to understand the state and local regulations, codes, standards, or laws that come into effect. Even then, you have to satisfy whoever is permitting the effort.

In Austin, the AHJ is the Austin Fire Department, with the Austin Fire Department Prevention Division being tasked with the permitting and inspections. Like many locations, this division within the department spends relatively little of their time dealing with flame effects. Sometimes they are very familiar with the needs and requirements of flame effects; other times they are trying to apply general requirements to something unique.

One of their requirements (a common one) is a flame effect plan for every flame effect that is being permitted. This document generally contains at least the following:

- The name of the person, group, or organization responsible for the production

- The name of the effect operator

- The dates and times of the production

- The location of the production

- The flame effect classification

- A site plan

- A narrative description of the flame effect

- The area affected by the flame effect device

- The location of the audience

- The fuels used and their estimated consumption

- Sources and flow of air for combustion and ventilation for indoor effects

- Flammable materials piping

- Storage and holding areas and their capacities

- Supplemental fire protection features

- Emergency response procedures

- Means of egress

- A current material safety data sheet (MSDS) for the materials (fuels) consumed in the flame effect

- Documentation that the combustible materials used for construction of the flame effects have been rendered flame retardant (generally only needed if cloth or other normally flammable material is in use)

Not all of these requirements are relevant for every effect, but most are. Under NFPA-160: *Standard for the Use of Flame Effects Before an Audience*, even a lit candle on a stage for a play is a flame effect and needs a plan and permit.

Clear diagrams are required that show:

- The site layout and location of the effect

- The exclusion zone around the effect where the audience is not allowed (usually at least a 15' radius around flame and 5' around fuel or controls)

- Locations of fire extinguishers and the fire watch personnel who will use them if needed

- Egress (escape) paths for the audience

A preinspection of the effects is often required. This doesn't have to occur at the location of the event, but usually the AHJ will want to see the equipment and watch you ignite it. It is always a tremendous help to have an "operating manual" for your effect when your effect is being inspected. Handing a printed copy of the manual to the inspector will win you big points. Your operating manual should include:

- A schematic of the effect

- Components

- Effect clearance plan

- Setup and tear-down instructions

- Test and inspection steps (leak-testing, visual verification of components)

- Operating steps

- Emergency shutdown procedures

I'll offer some tips about how to interact with the AHJ. In many cases, you will also have to interact with the permit holder for the event. In Texas, this is a licensed flame effect operator (FEO). Both the AHJ and FEO have the ability to add requirements beyond the ones in the codes, standards, or laws. You may find yourself frustrated by hoops they make you jump through.

I've never seen someone successfully argue with the AHJ or FEO. Even in situations where the person knew more about the technicalities than the AHJ or FEO, they didn't manage to "win" an argument about permitting. Both the AHJ and FEO are charged with the legal liability for the effect. If they feel that the effect operator is difficult to work with, they have to decide if they feel like putting their jobs, licenses, or people's lives on the line for someone who they may not trust. As an FEO, I've been told directly by the AHJ that it's my call and I can refuse to allow people to light up, even if I just don't feel good about it.

So, my tips are to make it easy for the AHJ and FEO to say yes. Keep your situation tight, provide abundant documentation, stay friendly and positive, and don't argue. The hardest tip I will offer is that you must look like you're prepared *not* to light up if they will not approve.

If they say no, say that you understand, and politely ask what could be done to get approval. Ask if there are alternatives that they may be aware of. Many times, I've seen a shift of attitude where the AHJ realizes that the effect operator isn't going to fight them about something and then all of a sudden becomes really helpful. But they usually start by acting reserved until they have a reason to trust. Their job is to protect lives and property not entertain people, so respect their needs if you want them to respect yours.

Enhancements

The degree to which you can enhance a flame effect like the boosh is unlimited. There are enhancements in all the areas described in the sections at the beginning of the chapter: "Fuel," "Control," and "Ignition." There are also plumbing enhancements and artistic enhancements. You can build flame effects into sculptures, mount them on vehicles, control them with hand gestures; there really is no limit.

I'll describe a few enhancements that are fairly common and easy to implement. There are additional enhancements in the following chapters on pilots and Arduino control, as well. The final chapter will provide some ideas about how far enhancements can really go.

Hose Protection

A 10' or longer run of propane hose sitting on the ground is easy for someone to step on. I usually take a 10' stick of 1½" or larger PVC pipe with me to boosh. I place the hose in it before hooking it up to the boosh. Gray PVC conduit also works well, as will any kind of tube or pipe that provides a protective shell.

Plugs and Caps

Since my booshes don't get as much use as I'd like, I have to leave them stored for periods of time. One of the things that has improved their state when coming out of storage is the addition of plugs and caps on all inlets and outlets. You don't have to use all of them, but here are the ones I use:

- Gas inlet: ⅜″ FFL brass cap

- Top of solenoid: ¾″ black iron male plug

- Low-pressure regulator hose: ⅜″ MFL brass plug

- Bottom of main effect vent pipe: ¾″ black iron pipe female cap

- Bottom of pilot cutoff valve: ⅜″ FFL brass cap

- End of high-pressure source hose: ⅜″ MFL brass plug

Except for the brass (which I can't ever bring myself to paint), I paint all caps or plugs bright yellow so that I can find them in my gig bag and distinguish them on the effect.

Quick Connects

Flared connections are dramatically better than taped joints for locations where you want to easily assemble and disassemble your device. Nevertheless, modifying flared joints requires gas to be shut off to that joint, a wrench, and appropriate tightening. If you've ever worked with air tools, you've probably encountered quick connect fittings that allow you to attach and detach components without tools, even while they're under pressure.

Quick connect fittings were described in Chapter 2. Hopefully, you noted the distinctions between quick connects rated for air (never to be used with propane) low-pressure propane, and high-pressure propane. Low-pressure propane quick connects are easy to mistake for high-pressure connects. But it is a huge bummer to do so. The low-pressure connect really isn't rated for high pressure and if used in a high-pressure system, will blow out in the first use, ruining itself forever. High- or low-pressure propane-rated quick connects are way too expensive to throw away like that. Be careful and make sure you're buying quick connects rated for at least 250 psi.

These fittings typically have ¼" MIP/FIP threaded connectors and can be used at any point in the system where mechanical strength is not required. Locating one at the end of the feed hose to the boosh is a common and useful design. This allows the gas feed to quickly and easily be removed.

It is also possible to buy hoses with premounted quick connect fittings. Consider this (and the necessary additional fittings to mate with the hose) when you're making your purchases. I generally want my hose to be more versatile and do not buy it with preinstalled quick connect fittings, but this can be a good option for some designs.

Alternate Feeds

If you have a lot of fun with booshes, you will eventually find yourself standing around staring at a propane cylinder covered in about 2" of ice and producing less gas than you could burp. There are a host of ways to get around this, and most are beyond the scope of this entry-level book. However, one method is fairly easy for all levels of enthusiast: additional feeds.

The plumbing tree of the boosh in this project has four branches. You can use a tee to plumb in a second gas inlet branch and add a second high-pressure source. You can alternate which source feeds the effect with the source's ball valve cutoffs. You'll still eventually freeze both cylinders, but it will take longer.

Switching back and forth between two sources and buying a complete second set of source gear (cylinder, regulator, ball valve, hose, etc.) is more complex (and expensive) than many people will be happy with, but it's straightforward and it works.

There are methods to manifold cylinders together before the regulator, but this is a dangerous undertaking that requires experience. Please don't try things like this until you've gotten really comfortable working with basic systems.

10

Pilot System

THE BOOSH PROJECT PRESENTED a simple, effective manual pilot system. That pilot required an external ignition source to get started and, once lit, stayed lit continuously. That's a workable approach, but when the wind blows the pilot out or you find you can't reach the pilot easily to relight it, frustration mounts.

To ignite the low-pressure gas jets of the pilot, we need heat above 878°F (470°C). There are two reasonable ways to achieve this that we can control remotely: a spark or a heater element. We'll discuss a couple simple spark mechanisms and describe the addition of a heater element to the boosh project.

Parts

Lighting a BBQ grill is a similar, if not identical, challenge to the one we're working on. In that system, there is also low-pressure gas that needs to be lit, ideally without waving a flame around. Numerous commercial solutions have emerged to address this problem.

Parts:

REF	ITEM	QTY
P1	12V DC mini hot surface ignitor	1
P2	Lamp cord 12′	1
P3	Corner brace 1″	1
P4	Bolt #10 × ½″	1
P5	Washer #10	2
P6	Lock washer #10	1
P7	Nut #10	1
P8	Momentary contact switch	1
P9	Heat shrink tubing (assorted sizes)	1
P10	16-gauge wire × 10″	1
P11	Two-pin trailer connecter extension	2
P12	Zip ties	4
P13	Hose clamp ½″	2
P14	Wire nuts	2

For the most part, these solutions fall into two camps: piezo-electric lighters and spark generators. Piezoelectric lighters are extremely common; the fireplace lighters recommended in other chapters of this book are generally piezoelectric lighters. *Piezo* (usually pronounced *pee-ay-zo* in the US and *pay-zo* in the UK) lighters apply stress to a dielectric material with a crystalline structure, which transforms mechanical stress to electric charge (and vice versa). Piezo materials can easily generate kilovolts, which is more than enough to generate a spark. Piezo wired the other way (receiving electricity and generating motion) can produce sound and is commonly used as a small buzzer or speaker source.

The downside to mechanical piezoelectric ignitors is that they can only support a limited wire run between the spark source and the electrode. They'll support a run long enough that you could mount the electrode at the pilot ring and the piezo controller near the needle valve, but trying to run the wire 10–15′ to the controller box would be difficult due to the voltage drop in the wire.

Other products exist on the market. Small electric spark units usually run on AA batteries, but they also have limits on how long a wire run they can sustain to the electrode. (See Figure 10-1.)

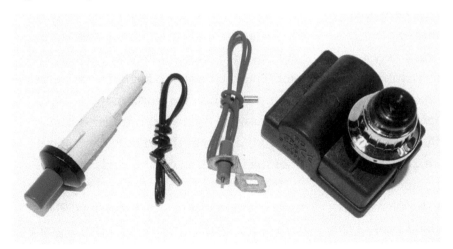

FIGURE 10-1: Piezoelectric and electronic ignitors

With some ingenuity, these units can be hacked to allow remote triggering, so that the power to the unit travels over the long wire. However, doing this is tedious compared to the alternative we're going to build in this chapter.

Construction

The 12V DC mini hot surface igniter is a smaller version of the instrument used to light gas appliances. (See Figure 10-2.) These igniters come in a wide range of voltages, AC and DC, and with differing lengths. The 12V unit is useful for our needs since we're already working with a 12V circuit. It also turns out that the speed with which the ignitors reach their target temperature is related to how long they are. The mini unit heats up to 1300°C (2372°F) in less than three seconds. The larger units draw more power, but these small units typically draw 20 watts. This works well for our needs.

Availability of the mini 12V units seems to ebb and flow in the market. There are periods when they are difficult and expensive to find; then it seems like they're widely available again. At the time of this writing, they are available online at sites like SparkFun for about $20 USD. Other vendors, such as Crystal Technica, seem to carry them more consistently, but at a higher price.

One of the potential downsides to the hot surface igniter is that the heater element is breakable. If subjected to harsh forces, it can crack or chip. Be careful with your igniter

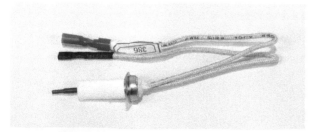

FIGURE 10-2: 12V DC mini hot surface igniter

and the pilot system once it's installed so it doesn't become damaged.

1. Wire the Igniter

Depending on where you purchased the igniter, it will have a variety of electrical connections, such as single connecter, two-spade connectors, and so on. If you're able to create mates for the ends by crimping or other methods, that's ideal. But if not, we can refit the igniter to use the same two-pin trailer connector we used on the boosh project.

Cut the existing connectors off (as close to the connectors as you can). Strip away about a ¼″ of the insulation off the wires. If your igniter has braided fiberglass (or possibly Kevlar) insulation, you may want to use a very tiny bit of superglue or nail polish (use a toothpick to apply) to keep it from unravelling further while we work on it.

Cut one of the two-pin trailer connectors (P11) in two at the middle. Trim about ½″ of the insulation off the leads of one of the connectors. Slip a 1″ (2.5 cm) segment of small-diameter heat shrink tubing (P9) over each of the leads. Slip a 1 ½″ (4 cm) segment of larger heat shrink tubing around both of the leads.

Twist the exposed wire from one of the connector leads around one of the leads from the igniter. The igniter does not have polarized leads, so it doesn't matter which wire attaches to which. You only need enough of a twist to hold the wires in place while you solder. (See Figure 10-3.)

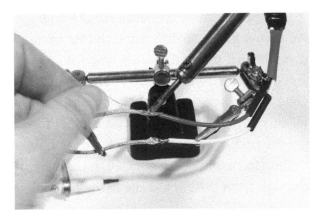

FIGURE 10-3: **Soldering the connector**

Solder the igniter leads to the leads of the connector. Cover each single lead with the small heat shrink and use a heat source (heat gun or lighter) to shrink. Then slide the larger heat shrink over both leads, and shrink it. (See Figure 10-4.)

FIGURE 10-4: **The completed connector**

2. Mount the Igniter

The igniter comes with a metal tab attached to the ceramic insulator below the heating element. Unfortunately, for our purposes, it's not very convenient. We want the element of the igniter to be close to a pilot jet so that it will ignite the escaping propane. We also want the body of the igniter to be below the pilot ring so that we don't continuously bathe the ignitor body in flame once the pilot is lit. (The ignitor is designed to work in a high-heat environment, but it will last longer if we subject it to less abuse.)

The problem is that the igniter's tab is perpendicular to the igniter body, so we need something parallel to the body that will allow us to attach the igniter to the vertical shaft of the end of the pilot tube. We could try bending the tab, but this is an easy way to crack the ceramic insulation if it's not done very carefully. Instead we'll fabricate a simple mount.

We will bolt the tab on the igniter to a 1″ 90° corner brace (P3) and use a ½″ hose clamp (P13) to attach the brace to the vertical copper tube. Place a ½″ washer on a #10 × ½″ bolt (P4), then pass the bolt through the tab and the brace. Place another ½″ washer, lock washer, and nut (P5, P6, P7) onto the end of the bolt and tighten. (See Figure 10-5.)

3. Wire the Pilot Cable

The cable is identical to the effect cable we made for the boosh in the previous chapter. It's a 12′ length of lamp cord (P2) with two-pin trailer connectors soldered on both ends.

FIGURE 10-5: **The mounted igniter**

Take a look at the instructions in that chapter for details if you need them.

4. Modify the Controller

We will be adding another, and this is important, *momentary contact switch* to the controller. You don't want to leave the igniter powered up any longer than you have to, because it gets very hot and very brittle. We can put the switch anywhere it will fit, and it could be any size switch you like. I used an arcade button switch (P8). This switch needed a 1⅛" hole, which I positioned ¾" from the top and ¾" from the left of the controller's surface. (See Figure 10-6.)

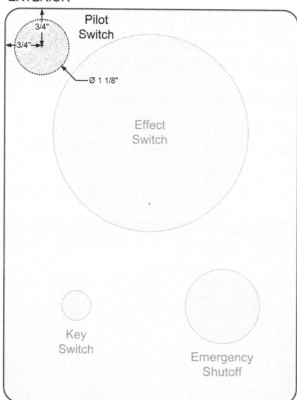

CASE EXTERIOR

Pilot Switch

3/4"

3/4"

Ø 1 1/8"

Effect Switch

Key Switch

Emergency Shutoff

FIGURE 10-6: Pilot switch positioning

I used my trusty step drill for the hole. It's always easier to drill through thin metal if you have it pressed against wood. A short piece of scrap 2×4, stood on end, provides a good back when drilling the mounting hole for this switch. For a cleaner hole, don't forget the trick of switching which side you're drilling from with each step. Push the switch through from the back and use the hex nut to secure it.

We'll need to bring in another two-pin connector from the outside. Hold the wires from the existing main effect connector hard to one side and slip the two wires from the pilot connector through the same grommet (one at a time). Tie them together in an overhand knot to provide strain relief. Mark the connector with a magic marker, tape, or some other mechanism to be able to distinguish it from the effect connector.

The wiring goes as follows:

A. Unscrew the wire nut with the black (negative) wires, add the black (or possibly white) wire from the pilot connector, and twist the wire nut back onto them. If you used heavy gauge wire, then you might have to step up to a larger size wire nut.

B. Solder a short length of 16-gauge wire (P10) to the COM terminal of the pilot switch. You'll need about 2" (5 cm). Use a wire nut (P14) to connect this wire to the red lead from the pilot connector.

C. Solder a 6" (15 cm) onto the NO terminal of the pilot switch.

D. Cut the wire that goes from the effect switch to the emergency shutoff switch in the middle and strip ½" (1 cm) from each end.

E. Use a wire nut (P14) to connect the wires from the previous step to the wire from the pilot switch

(See Figure 10-7.)

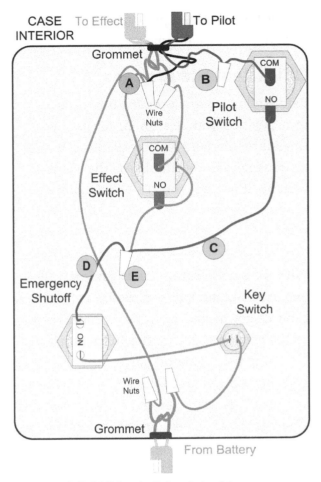

FIGURE 10-7: Additional pilot switch wiring

Assembly and Testing

The steps below assume the boosh is set up for operation per the instructions in the previous chapter. The exception is, as noted, that the accumulator cutoff is closed and the effect cable is not connected. You may have noticed that I just repeated myself. That's excellent; it means you're paying attention and hopefully recognize how important it is to not fire the boosh accidentally!

Connect the Cable

Connect the pilot cable to the connector and the igniter. I highly recommend adding zip ties (P12) around the pilot tube to hold the cable in place.

To be honest, once I've connected the cable to the igniter and zip tied it to the pilot tube all the way down to the needle valve, I don't disconnect it again. When I break down the system, I coil the pilot cable and keep it with the pilot/main tube assembly.

Quick Test the Igniter

Before allowing any propane into the system, position yourself where you can see the igniter element and press the button for a few seconds. You should see the element heat up. If not, see the "Troubleshooting" section.

Arm the Pilot

With the source cutoff, effect gas inlet, pilot and accumulator valves closed, and the effect cable disconnected, open the cylinder valve all the way and back it off a quarter turn.

Using the regulator gauge, open the high-pressure source regulator somewhere between 5–10 psi. Open the source cutoff and effect gas inlet valves. Open the needle valve all the way, but keep the pilot cutoff closed.

Turn the key switch to the on position and verify that the emergency cutoff switch is disengaged (on).

Test the Pilot

Open the pilot valve cutoff and press the pilot switch on the controller for 5–10 seconds. When the pilot ignites, (shown in Figure 10-8) release the pilot switch.

If the pilot does not ignite, try the switch one more time. If that doesn't work, go to the "Troubleshooting" section.

Troubleshooting

Before troubleshooting, make sure all ball valves on the system are closed.

FIGURE 10-8: The pilot lit

Quick Test Fail

If the quick test failed, you will need to test the output of the pilot cable by unplugging it from the igniter. Use a multimeter to verify that the cable's connector is producing 12V when the pilot button is pushed.

If the pilot cable's connector does not produce 12V when the pilot switch is pressed, first make sure that you're connecting the pilot cable to the pilot connector on the controller. If you're using the correct connector, then unplug the pilot cable and test the output of the controller's pilot connector when you press the pilot switch.

If you don't see 12V on the pilot connector, you'll need to open the controller and check the connections at each of the wire nuts that the pilot switch wires are connected to. You'll also need to check the connection to the emergency cutoff switch.

Pilot Light Test Fail

If the pilot didn't light, close the pilot cutoff valve. Wait a couple minutes and try the igniter quick test again (visually watch the igniter to get red hot). If you see the igniter heating up, but the pilot did not light, the igniter is probably positioned so that it isn't getting any propane.

Attempt to reposition the igniter (swing it around the tube, move it higher or lower) so that it is in the path of one of the jets on the pilot ring.

Operation

Operation is very simple. Follow the operating instructions for the boosh in the previous chapter, but use the pilot switch to light the pilot instead of a lighter or torch. You now have the added benefit of being able to relight the effect if the pilot blows out.

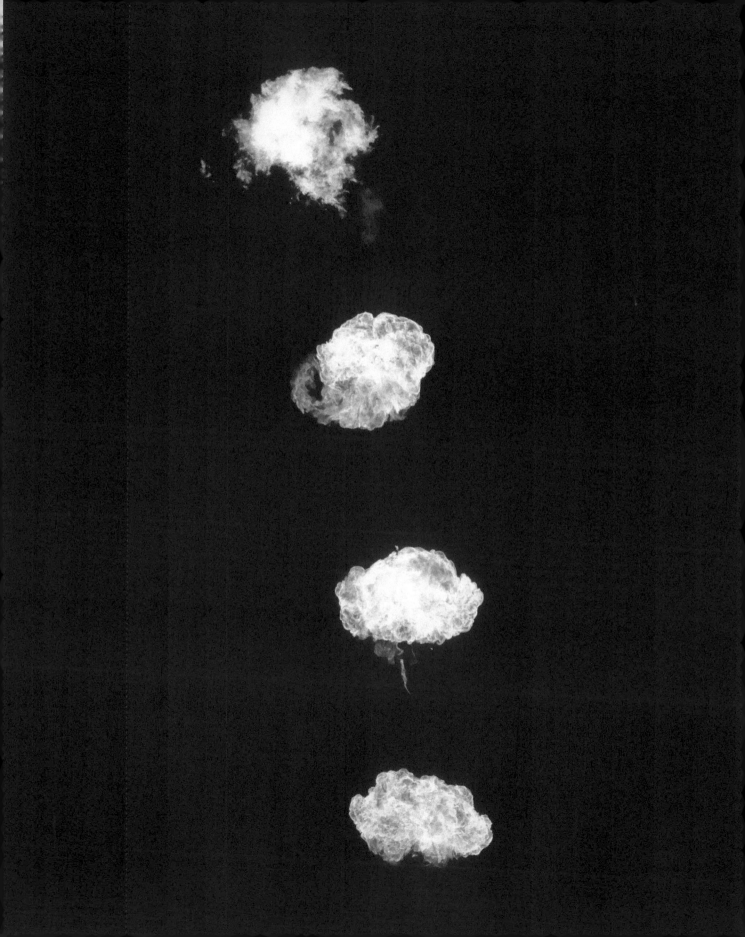

Arduino Control

I'M THE FIRST PERSON to get excited about the thrill of creating a huge fireball in the sky. But once you've built your boosh and are looking for something more, what's next? For many, this will involve bigger valves and higher pressures, but I've always been interested in other dimensions of flame effects.

The solenoid valve we've used for the boosh turns on and off, and it opens and closes. That may sound limited, but it's also binary. Manipulating binary values is the heart of the computer revolution, so clearly there's a world of things you can do with 1s and 0s.

The possibilities, once your boosh is under programmatic control, are limited only by your imagination. Feedback and timing offer a rich palette of interactive and patterned flame effects. There are digital thermometers, pressure gauges, and scales that can provide information about how big a boosh is possible at any given moment. Proximity and facial recognition sensors, force gauges, and microphones can trigger flames as responses to anything you can instrument.

In this chapter, I want to entice you with a couple of the more basic things to do in this huge playground of ideas. Nevertheless, safety remains our primary concern and there are additional concerns once computers are involved. It's also important to discuss where computer flame control crosses the line into show control.

Safety Note

We touched on the topic of show control systems in Chapter 9, "The Boosh." NFPA-160 *Standard for Flame Effects Before an Audience*, states that main show control can only be part of Group V and VI effects, which have more stringent requirements than Group III, the classification most enthusiast-built flame effects fall into. NFPA-160 does not provide a definition of main show control even though it uses the term as a distinguishing factor between flame effect groups. There are terms such as *main show control* and *life safety controller* that are used with the expectation that the reader is working in an industry where these are terms of art and well-understood.

The companies that build and operate professional show control systems use special purpose-built controllers that cost hundreds or thousands of dollars; they do not use off-the-shelf consumer microcontrollers like Arduinos or equivalents. There are good design reasons for this, involving redundancy and certification. They charge a lot of money and do amazing things. This book is in no way suggesting that enthusiast-built effects are the equivalent of what you see in Las Vegas or big stadium concerts. If you're interested in the differences between the professional and hobbyist tiers of flame effects, do your homework; don't use the DIY motto of "How hard can it be?" to avoid learning something fascinating!

To keep within the bounds of Group III effects and to stay safe, use the following guidelines as a starting point. Understand, as with all aspects of working with an authority having jurisdiction (AHJ), that even with these constraints, the inspector or licensed flame effect operator (FEO) may decide that they aren't okay with your effect. Be prepared to demonstrate and display exactly why and how your device is safe and always be ready to shut down, or not even set up, if the AHJ or FEO tells you to stop.

- Always design your system so that you have direct visual observation of the pilot and flame.

- Always include a "dead man's switch." This is a normally open momentary contact switch that makes it impossible for the computer or microcontroller to fire the effect unless the operator is holding the safety switch down (this is in addition to a fire switch that signals the controller to fire the effect).

- Always include an immediately accessible emergency shutoff switch that cuts power to all aspects of the system with a single motion.

- Always use solenoid valves that *fail-safe*, meaning that if they lose power, they close themselves (aka, *normally closed*).

- Always include at least two quarter-turn fuel shutoff valves in the effect path so that you can shut the fuel off completely with a single motion from your operating position (and at the effect).

- Never operate your effect with people in proximity to the effect (within the a 15–20′ exclusion zone).

Having laid out these safety concerns, many things are still possible with a microcontroller like the Arduino or equivalent and a flame effect. This chapter will focus on adding a new mode: patterned fire.

To be more specific, we will build a circuit on an Arduino protoshield that will, while the fire and safety buttons are pressed, open and close the effect valve using an adjustable duty cycle. This provides a sequence of fireballs of adjustable size (based on duration) and interval. We'll add the ability to switch between automatic and manual operation modes and use LEDs to provide visual control signal feedback from the microcontroller.

Since it underlies the core of this approach, let's dive a little deeper into the idea of a duty cycle. In short, a *duty cycle* describes the ratio of on time to off time for a signal. A 10% duty cycle means that the signal is on 10% of the time and off for 90% of the time. A 50% duty cycle means the signal is on for the same amount of time it is off. (See Figure 11-1.)

You'll notice that I haven't said anything about how long the cycle runs, just the ratio of on to off. A 10% duty cycle could mean that the signal is on for 1 second and off for 9 seconds or it could mean that it's on for 20 minutes and off for 180 minutes. To usefully control the boosh using a duty cycle, we'll also

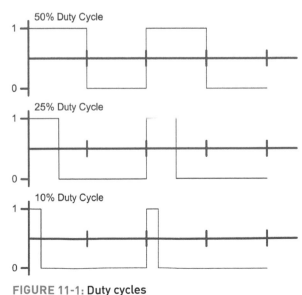

FIGURE 11-1: Duty cycles

have to control the wavelength of the signal (the length of time of the on segment plus the off segment).

Generating a signal like this doesn't necessarily take a microcontroller. You could build a circuit with the venerable 555 timer to do this. The Arduino or equivalent may not be critical, but it offers some serious advantages. Managing the timing of signals and using the logic of software is much more flexible than doing it all with hardware. As a platform for adding functionality, microcontrollers are unbeatable.

I'm not going to go so far as to describe the case and the buttons and all the other goodies that turn this microcontroller/shield into a full boosh control box. If you've built the previous projects, that part will be easy. More importantly, I expect this project to be a foundation and that you will modify and change it well beyond a basic capabilities provided. By the time you settle on the way you really want to implement this, your interface and enclosure are likely to be very different than anything I'd suggest.

Parts

Parts:

REF	ITEM	QTY
MC1	Arduino Uno (or any Arduino IDE programmable board usable with standard shields)	1
PB1	ProtoScrewShield kit (Maker Shed, SparkFun, etc.)	1
Q1	Logic level N-channel power MOSFET (see notes below)	1
D1	Diode (1N4001)	1
LED1	3.2-3.4 VDROP LED	1
LED2	2.2-2.4 VDROP LED	1
POT1, POT2	Potentiometer (10K ohm)	2
R5	Resistor (150 ohm)	1
R4	Resistor (330 ohm)	1

(continues)

REF	ITEM	QTY
R3	Resistor (470 ohm)	1
R1, R2	Resistor (1K ohm)	2
S1, S2	Momentary contact switch, normally open	2
S3	SPDT switch	1
W	22-gauge solid core wire (multi-color optional)	1

For a complete system, you will also need the parts listed for the control box of the boosh project.

Figure 11-2 shows a schematic diagram of the project with onboard and offboard components.

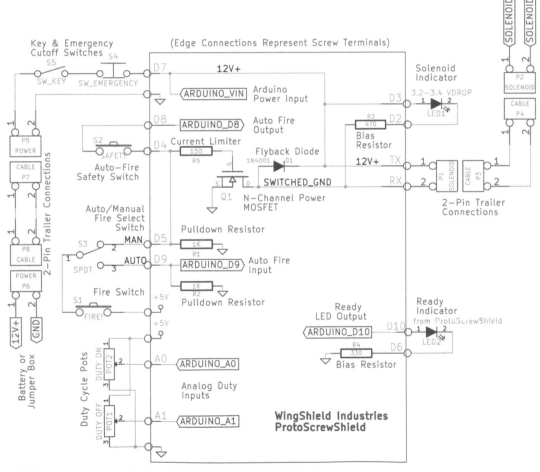

FIGURE 11-2: Schematic diagram of the boosh shield

Design

While this is an extremely simple circuit, there are a few design points worth discussing. Some of the approaches being used may not be familiar to readers, so we'll spend a little time explaining each of the key design elements in this controller.

Powering the Arduino

The Arduino Uno, or equivalent, is powered by either 5V directly or 7–12V on the VIN pin. Since we have a handy source of 12V in our boosh power supply, we can route it to the VIN pin directly on the ProtoScrewShield. This means that the Arduino will power up or down with the state of the key and emergency cutoff switches. The Arduino powers up very quickly and the ready LED will give us four blinks to let us know it's happy and ready to go, then start blinking in time with the duty cycle.

The MOSFET

There are many ways for an electronic circuit to *switch a load*. By this, I mean to control the flow of electricity to something that consumes electricity. To determine which approach to take, it's essential to understand what that something is. Most importantly, we must know how big the load is, frequently measured in amps. You also have to know things like whether the load is intended for alternating current (AC) or direct current (DC.) Switching an LED that draws milliamps is very different than switching a big AC motor on or off. This topic is big enough

that there are graduate degree programs in electrical engineering that prepare people to spend careers exploring the nuances of resistive, inductive, and capacitive loads and how to switch them.

We'll stay on the hobbyist end of the spectrum and concentrate on how to get a microcontroller with a 5V output to open and close a 2-amp solenoid valve. Even this topic offers a wide range of choices involving relays, solid-state relays, transistors, JFETs, MOSFETs, and other semiconductor solutions. There are fantastic books that will help you come to terms with these options (look for books by Charles Platt, Paul Scherz, Gordon McComb, and Forrest Mims, among others). Over the years, I've settled on MOSFETs for problems like this, so I'll concentrate on explaining that approach. (See Figure 11-3.)

MOSFETs allow a low voltage and current to control, or *switch*, a higher voltage and current. N-channel MOSFETs use a positive voltage at the gate to let current flow between the source and the drain (aka switching the circuit on). There are a wide variety of logic level N-channel MOSFETs on the market that will meet the needs of this project. The following chart lists five commonly available parts, any

FIGURE 11-3: FQP30N06L MOSFET

of which will work. While the data sheets don't always provide identical information, I've tried to normalize the values.

PART	MANUFACTURER	MAX R_{DS} (ON)	V_{DS}	I_D (CONTINU-OUS @ 25°C)
IRLB8721	International Rectifier	(V_{GS} 4.5V, I_D 25A) 0.016 ohm	30V	62A (V_{GS} 10V)
IRL520	Vishay	(V_{GS} 5V, I_D 5.5A) 0.27 ohm	100V	9.2A (V_{GS} 5V)
RFP12N10L	Fairchild Semiconductor	(V_{GS} 5V, I_D 12A) 0.20 ohm	100V	12A (V_{GS} 5V)
FQP30N06L	Fairchild Semiconductor	(V_{GS} 5V, I_D 16A) 0.045 ohm	60V	32A (V_{GS} 10V)
RFP30N06LE	Fairchild Semiconductor	(V_{GS} 5V, ID 30A) 0.047 ohm	60V	30A (V_{GS} 5V)

For this project, we want a MOSFET that can be driven directly from the output pin of the Arduino, which operates at 5V. We would state this as a 5V gate voltage (V_{GS}) requirement. MOSFETs that require at least 10V V_{GS} are more common and require a transistor or some other intermediate switch between them and the microcontroller's 5V output. Most MOSFETs will work at a variety of gate voltages, and varying the gate voltage allows for analog uses of MOSFETs such as in amplifiers.

Though the necessity is hotly discussed online, with both camps making strong points, the current limiting resistor (R5) that guards the gate of the MOSFET is there to make sure that the Arduino pin does not exceed its stated 40 milliampere capacity. The MOSFET shouldn't draw nearly that much, but a resistor is cheaper than a new Arduino, so it's cost-effective insurance.

The source on the MOSFET is connected to ground (GND). This means that loads, such as our solenoid, are permanently connected to the positive voltage and have their ground turned on and off to switch them. This is counterintuitive for many people and has important implications if your project uses a negative chassis ground. If the body of the project can provide a ground path, you have to be extra cautious not to create a short that will provide a path to ground for the load.

MOSFETs dissipate heat when handling loads. The TO-220 package, which all of the above MOSFETs are available in, has a metal tab that can be attached to a heatsink to allow it to handle loads at the high end of its capability. A single 2-amp solenoid should not require a heatsink unless you're holding it open for long periods (which is hopefully not the case with a boosh).

MOSFETs are static sensitive, so be careful not to fry one with static build-up. Ground yourself by touching a grounded metal object like a water pipe or large appliance before handling. Try to avoid handling the leads of the MOSFET, and don't let your clothing touch it. If you work on electronic projects frequently, it's helpful to construct a grounded workbench and use an antistatic wrist guard.

The Flyback Diode

A solenoid, like the valve we're using, is an electromagnet. An electromagnet is a coil of wire that is electrified, creating a magnetic field. This field magnetizes a metal rod and pulls it into the coil. The rod is attached to the valve and the valve opens. When the electricity is cut off, a spring returns the rod to its original position, closing the valve.

Electrified coils of wire are amazing things. Without attempting to go into the wonderful world of impedance, reluctance, reactance, and inductance (all-important in the world of coils), I'll simplify and say that coils store energy. Having stored energy in the coil as a magnetic field, when you cut off the electricity, the magnetic field collapses and releases the energy (aka *current*) involved. This surge of current has to go somewhere, and if it heads back into our circuit it can possibly damage components.

The easiest way to protect against this is to include a diode across the leads of the load. The diode allows a quick path for the current to flow from the coil back into itself so that it rapidly dissipates rather than trying to find a path back through the rest of the circuit. This is referred to as a *flyback, flywheel, freewheel, suppressor, suppression, clamp,* or *catch* diode. I like flyback.

Any diode that exceeds the current and voltage characteristics of the coil will work. The 1N4001 in our circuit (D1) is actually overkill (and not the fastest choice), but it's easy and cheap to acquire. Other 1N400x parts will also work, as will a range of other diodes. The reverse voltage rating on the diode should be at least the voltage applied to the coil (12V in our case), but more is better. The current rating (or at least the surge rating) must also exceed the load (about 2 amps in our case.) The 1N4001 has a 35V reverse voltage and a peak forward surge current of 30 amps.

Pullups and Pulldowns

Pullup and pulldown resistors force the default state of a microcontroller pin to a known value. A pullup resistor is placed between the pin and V_{DD} (5V). A pulldown resistor connects the pin to V_{SS} (ground). When a signal is applied to the pin, it provides a less resistant path than the one with the resistor and the pin gets the signal value. Without the signal present, the pullup or pulldown forces the pin to the selected 1 or 0.

The Arduino Uno has the ability to apply internal pullup resistors to the input pins via software. This means we don't have to add physical resistors to the circuit to force the default state on input pins if we want that default to be 1 (aka *true* or 5V). However, forcing the state to default to 0 (*false*, ground) requires an external pulldown resistor.

Our circuit requires two pulldown resistors (R1 and R2). One is in the path of the D8 output, which will fire the boosh. This means that the default state of the gate of the MOSFET will be closed.

The other pulldown is on the D9 input. We're going to include the ability to use the fire button for either auto or manual fire. This means that the fire button has to provide 5V to the MOSFET in manual mode to open it and fire the boosh. In auto mode, the fire button provides an input signal to D9. Since we've wired the fire button to provide 5V, that means that we have to make the D9 input active on a positive input, so the default state should be ground.

Potentiometers as Analog Inputs

Using a potentiometer, or *pot*, as an analog input is a common technique. But, if you've never understood why they're useful, it's worth me taking a minute to describe their value. When I was young, I never really understood why potentiometers have three leads. It seemed like they only needed two, one on either end of the variable resistor. I imagined that it had something to do with which way you wanted to turn the knob.

But the actual reason there are three leads is the very reason that pots make great analog inputs. It's because a potentiometer is also a voltage divider.

A voltage divider is a simple circuit. You take two resistors in series with a positive voltage on one end (V_{IN}) and ground, or negative voltage, on the other, and measure what the voltage is in between them. (See Figure 11-4.)

The in-between voltage, V_{OUT}, is based on the ratio of the resistor on the ground side (R_2) to the combined resistance of both resistors ($R_1 + R_2$ or R_{EQ}). If V_{IN} is 5V and both resistors are the same size (R_2 is 50% of R_{EQ}), then V_{OUT} is 50% of V_{IN} or 2.5V. If R_2 is 20% of R_{EQ}, then V_{OUT} is 20% of V_{IN} or 1V. I'll leave it to the curious among you to prove this with Ohm's law.

A potentiometer is a resistor with a *tap*. The tap is a movable wire that touches the resistor somewhere along its length and provides a place to extract V_{OUT}. It essentially turns a single resistor into two resistors: R_1 and R_2 in series. If you measure the resistance of the pot across both outside leads, you get a fixed value; that's the single resistor, or R_{EQ}. If you measure between either of the outside leads and the middle lead, the tap, you get a variable value.

We turn the pot into a voltage divider by providing 5V on one lead and ground on the other and using the middle lead to measure V_{OUT}. This uses the pot as a variable voltage component instead of a variable resistance component. This variable voltage makes a perfect analog input. We'll use two of them (POT1 and POT2) to produce the inputs for our duty cycle lengths.

Safety or Dead Man's Switch

I mentioned in the safety section earlier that we needed to include a dead man's switch in the circuit. The term *dead man's switch* is a reference to a technique that requires a hand on a switch to keep something from happening. If the switch is released, something blows up or something shuts off or whatever is intended to be your fail state happens.

In our case, we need to have absolute confidence that some software glitch or bug

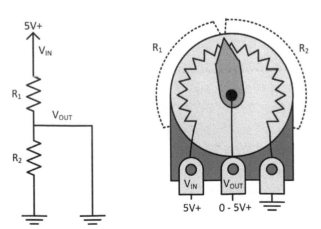

FIGURE 11-4: A simple voltage divider

doesn't accidentally open the solenoid valve that we've connected it to. We want our system to "fail safe" (as opposed to "fail deadly"). Computers are great until they aren't. And you don't want that moment to occur when it might cause harm or damage.

To add this kind of safety, we'll interrupt the path between the computer output pin and the MOSFET with a normally open momentary contact switch (S2). Unless you are holding this switch closed, the Arduino cannot switch the MOSFET on and open the solenoid valve.

Auto/Manual Fire Select

While pulsed operation using duty cycles is cool, sometimes you just want to hit a button and get a fireball. Sometimes ghosts get in the machine and the microcontroller chooses not to do your bidding, or something

breaks. It's always great to have a fallback that lets you use the controller even if the fancy stuff isn't working.

To allow operation as a manual boosh, the design includes a single pole double throw (SPDT) auto/manual selector switch (S3). *Single pole* means that you're switching a single wire as input. *Double throw* means it has two different outputs it can switch between. An on/off switch is a single pole single throw (SPST) switch, meaning it's either connected or disconnected from a single output. There are many other combinations: DPDT, SP4T, 4PDT, and so on. I have a couple 4P10T switches (four pole, ten throw) switches that are cherished vintage treasures.

Our SPDT switch routes the output of the fire button (S1) to either the Arduino D9 input for automatic mode, or directly to the gate of the MOSFET (Q1) for manual mode.

Construction

We'll be soldering components to the protoboard and doing *point-to-point* wiring. This involves cutting short lengths of wire and soldering them directly between components. The protoboard we're using makes this even easier, since a set of the holes on the board (which we'll refer to as *pads*) are electrically connected. We don't actually have to solder the wires to the components; we can solder them to nearby pads and the board will make the connection to the component for us. Other than breadboarding, this is the most common prototyping method in use these days.

Not very many years ago, I'd have recommended wire wrapping, but that appears to be a thing of the past. Not very long from now, I expect the ability to fabricate custom printed circuit boards (PCBs) at home, or order them for overnight delivery, will be trivial even for new makers.

Designing PC boards is easier right now than most people realize. Tools such as Fritzing have made PC board development cheap and easy. There are lots of places that will accept PC board design files online and send cheap boards in days or weeks. I'll give a shout-out to OSH Park (oshpark.com) as a great example of a site that combines your designs with others into big orders so that everyone gets low-priced boards. Figure 11-5 shows an example of one of my boosh controllers that OSH Park fabricated for me.

If you enjoy building electronics projects, consider stepping up to making your own boards. The excitement of seeing a PCB that you designed is a real thrill!

For this project, I wanted to limit the construction as much as possible, and the way to do that is to use an Arduino proto-shield. These are kits that provide a PCB and a few parts which, when constructed, attach to the Arduino and provide a place for you to add your own circuits. My favorite protoboard is the WingShield Industries ProtoScrewShield (wingshieldindustries.com). This is carried by multiple online sources and not only provides a prototyping space, but also adds screw terminals for all the Arduino pins. We'll be modifying the board slightly for our needs, but the combination of proto-space wiring and screw terminals means that we can build our boosh controller shield with a minimum of soldering.

You could construct this circuit on any protoboard you like with appropriate changes to the wiring. Given the safety concerns regarding fire, don't build this on a plastic breadboard. The chances of a part or wire falling out are too risky and, while the MOSFET we're using shouldn't get too hot, it has the potential to get hot enough to melt plastic.

Construct the ProtoScrewShield with Mods

The first step is to construct the ProtoScrewShield per the visual instructions at http://wingshieldindustries.com/products/proto-screwshield-for-arduino/ with a few modifications. (See Figure 11-6, next page.)

1. Do *not* solder the eight-pin stackable header into the RX, TX, D2-D7 section on the lower right part of the board (go ahead and solder all screw terminals).

2. The D13 LED and resistor are used elsewhere; do not solder them onto the board.

3. The reset is optional (but useful).

4. The A5/D19 switch is optional.

We could technically include fewer of the screw terminals, since we won't use all of them, but they offer support for future expansions, so it's worth including them all. The Blu-Tack adhesive tip for holding components while soldering on the WingShield site is one of the best tips I've encountered in years. If you haven't tried this method, I highly recommend it!

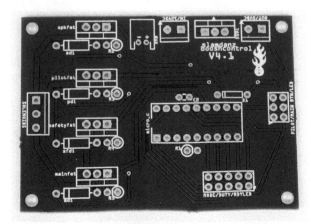

FIGURE 11-5: Boosh controller printed circuit

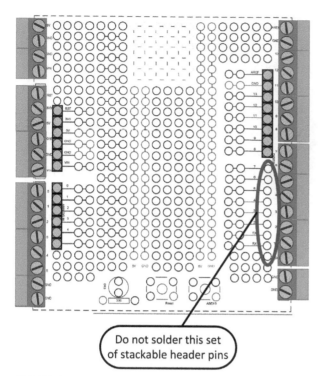

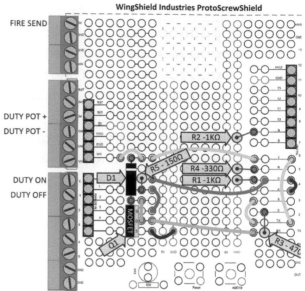

Do not solder this set
of stackable header pins

FIGURE 11-6: ProtoScrewShield with mods

FIGURE 11-7: Parts layout diagram

It's a good idea to quickly check your solder joints by using a continuity tester (or ohm meter) to verify a good resistance-free connection between the screw terminals and the header pins. On the section of RX-D7 where we did not solder the header pins, test against the labeled pads on the protoboard.

Parts Layout

For this project we'll solder seven components and ten wires onto the ProtoScrew-Shield. (see Figure 11-7.)

Dry-fit the parts loosely onto the proto-board before you begin soldering so that you're sure you have everything you need.

For the following two sections, I'll be using a notation based on the following diagram to describe where to place and solder components. (See Figure 11-8.)

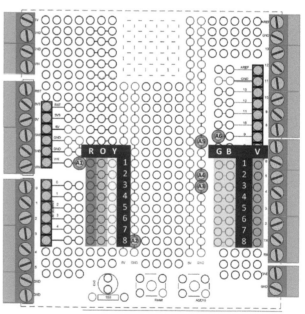

FIGURE 11-8: Parts placement legend

Solder the Onboard Components

There are five resistors, a MOSFET, and a diode to solder to the protoboard. As noted earlier, be careful handling the MOSFET to avoid static discharge, especially on its pins. While these are pretty robust parts, it's no fun to find out it's dead after you've soldered it in. Looking at the MOSFET from the front, the pin standard is, left to right, gate-drain-source. Note that the MOSFET is placed on the board with it facing left.

Diodes are polarized. They only allow current to flow in one direction, from positive to negative. (My apologies and respect to science, since electron flow is in the opposite direction. Ben Franklin errantly described the conventional flow direction and it stuck, so I'm using it here for simplicity's sake.) Most diodes have a line or ring marked on the cathode (-) side of the case. In our case, as a flyback diode, we're orienting it so that the cathode (-) attaches to the 12V+ part of the circuit. Be careful to orient it in this manner. This is counterintuitive to many makers since we're connecting the negative lead to the positive part of the circuit; usually the negative lead connects to the negative section of a circuit.

Resistors can be laid flat or stood upright. The upright stance often works well when using them in a tight space. The leads are generally long enough that you can stand one upright in a hole and route the other lead one or two holes over.

Once you've soldered in the components, flush cut the leads on the bottom of the protoboard. The table below lists the wiring points, and Figures 11-9 and 11-10 provide a topside view of where they will go.

PART #	PART	FIRST POINT	SECOND POINT
R2	1K ohm resistor	A5	A6
R4	330 ohm resistor	A4	G2
R1	1K ohm resistor	A3	G3
R3	470 ohm resistor	B6	B8
R5	150 ohm resistor	O3	O4
Q1	MOSFET gate	R6	
Q1	MOSFET drain	R7	
Q1	MOSFET source	R8	
D1	Diode cathode (line)	R1	
D1	Diode anode (no line)	R5	

Safety Note

On this project, I recommend that the lead on R3, which has to route over two holes between the connections "Solenoid LED –" and "Solenoid –," should be covered with a small section of heat shrink tubing. This will provide insulation as additional insurance along a possible activation path (the ground path) of the solenoid.

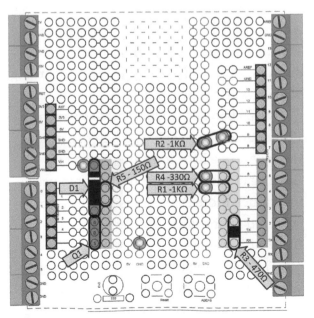

FIGURE 11-9: Component soldering diagram

FIGURE 11-10: **Completed component soldering**

WIRE NUMBER	FIRST POINT	SECOND POINT
1	A1	O1
2	Y1	G1
3	V1	V7
4	V3	V4
5	Y3	B3
6	G5	G7
7	Y4	Y6
8	O5	O7
9	Y7	G8
10	Y8	A2

Solder the Wires

There are 10 wires to solder for the circuit. You don't need to use different color wire, but it's always lovely to have color-coded wiring in a project. I will usually make an effort to have at least black ground wires and red power wires. If I have multiple voltage levels (as in this project) I'll use yellow as the alternate positive voltage.

Solid core wire is your friend when wiring protoboards. Solid core 22-gauge wire has a current capability of 5 amps and is available in multicolor packs from many online sources. This is also useful for breadboarding, so it's a worthy investment.

Strip about ½″ (10 mm) off each end of the section of wire being soldered. It's okay if the wire is a little long and doesn't sit flush against the board, but try not to use very much more than you need. Once you've soldered all the wires in place, cut the leads flush with the bottom of the protoboard.

The following table lists the points each wire will bridge and Figures 11-11 and 11-12 provide a topside view of where the wires go.

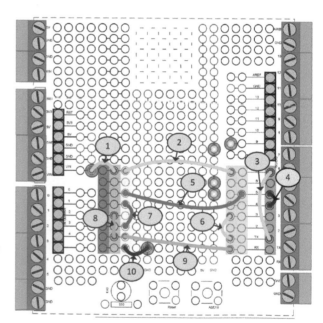

FIGURE 11-11: **Wire soldering diagram**

Solder the Outboard Components

There are a variety of components that will now attach to the ProtoScrewShield via the screw terminal connectors. These can be

FIGURE 11-12: ProtoScrewShield soldering complete

fabricated as independent parts using the appropriate components, with wire runs sufficient to meet the needs of the case you will be using.

Note that the ends of the wires that will be connecting into the screw terminals need to be stripped no farther than the depth of the terminal. You don't want bare wire emerging from the screw terminals.

Solenoid Indicator LED

This requires soldering two wires to the leads of the 30 milliamp LED. This is more durable if you cover the exposed leads and attached wires with heat shrink tubing. Don't forget to mark the polarity of at least one of the leads.

Ready Indicator LED

This should be treated the same as the solenoid indicator LED, but using the 20 milliamp LED.

Identifying LED Polarity

LEDs are light-emitting diodes; therefore, they are polarized. The ability to determine which lead from the LED is positive and which is negative is important. When looking at a non-surface mount LED, you can determine its polarity in one of three ways. The diagram in Figure 11-13 shows all three methods.

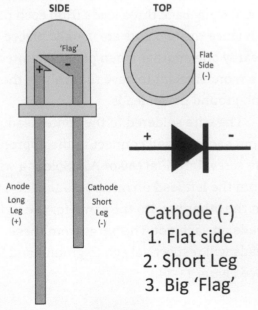

Cathode (-)
1. Flat side
2. Short Leg
3. Big 'Flag'

FIGURE 11-13: LED polarity

Solenoid Output Connector

This could be as simple as the wires from a two-pin trailer connector, as used in the previous chapters. You may also want to consider changing to a panel-mounted socket of some kind rather than bringing the leads in through a grommeted hole.

If you do use the leads from the two-pin connector itself, you may want to add some additional wire to them so that you have room to tie a knot inside the case for strain relief (see Chapter 9, "The Boosh.") Also, make sure to have the connector itself on the outside of the case before attaching to the screw terminals—an easy mistake to make!

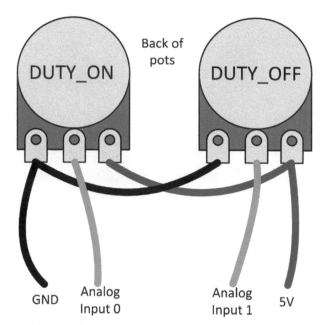

FIGURE 11-14: Potentiometer wiring

Duty On and Duty Off Potentiometers

We're bringing the output from two potentiometers configured as voltage dividers into the analog inputs on the microcontroller. We could bring back three leads from each pot, but since the two pots are usually placed nearby one another when panel mounted, it's more efficient to have them share the 5V and ground power leads.

The wire soldered to the center lead from each pot will connect to the appropriate screw terminal (A0 or A1). Solder a wire from the left lead on one pot to the left lead on the other pot. Do the same for the right leads on each pot. The wires from these leads will respectively go to ground and 5V. (See Figure 11-14.)

Fire Switch and Auto/Manual Select Switch

The fire and auto/manual select switches are wired in series. Solder a wire from one lead (COM) of the fire switch; this will go to a 5V screw terminal. Solder one end of a wire between the remaining lead (NO) of the fire switch and the central lead (COM) of the SPDT auto/manual select switch.

Solder wires to the two outside leads of the SPDT switch (often just labeled ON and ON). One of these will be the wire that selects manual fire and the other will select automatic fire. These should correspond to the switch position, but it never hurts to check with a multimeter. Connect these wires to the appropriate screw terminals.

Safety Switch

The safety switch requires a wire connected to each of the COM and NO leads. These will go into the appropriate screw terminals.

Emergency Cutoff and Key Switches, and Power Connector

The emergency cutoff and key switches were described in detail in Chapter 9, so please refer to it for specifics on these switches. Essentially, we will wire these two switches in series as part of the 12V+ input path. For a complete project, you will also need the components from the boosh controller, the various cable runs, the solenoid, the power source, and the pilot igniter (if it was used).

Figure 11-15 shows the set of required outboard components to complete this project. You may notice one of my favorite components in the picture, a combination emergency cutoff and key switch!

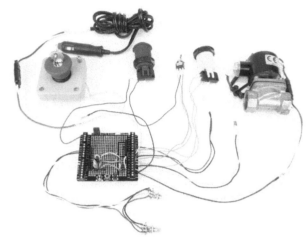

FIGURE 11-15: Outboard components

Software

The software to drive the controller is very simple. We can use the code's main loop as the mechanism to drive the duty cycle.

You'll need to download and install the Arduino environment. Use the standard setup, you won't need any special libraries or drivers for our sketch. Dozens of great tutorials are available online that show you how to get started with Arduino.

Start a new sketch and save it with whatever name you like, I saved mine as booshcontroller.ino. Select your board type and COM port.

Here's the complete code. We'll look at each section individually:

```
1    // Define pin names
2    #define FIRE_OUT 8
3    #define FIRE_IN 9
4    #define DUTY_ON 0
5    #define DUTY_OFF 1
6    #define READY_LED 10
```

```
7   #define scale_l 100
8   #define scale_h 2000
9
10  //---------------------------------------------------------
11  void setup() {
12
13    // Initialize pins
14    pinMode(FIRE_OUT, OUTPUT);
15    pinMode(FIRE_IN, INPUT);
16    pinMode(READY_LED, OUTPUT);
17
18    // set FIRE_OUT LOW (a good safety step)
19    digitalWrite(FIRE_OUT, LOW);
20
21    // 4 blinks of Ready LED as Power On Self-Test (POST)
22    for (int i = 0; i < 4; i++){
23     digitalWrite(READY_LED, HIGH);  // turn the LED on
24     delay(500);              // wait for a second
25     digitalWrite(READY_LED, LOW);   // turn the LED off
26     delay(500);               // wait for a second
27     }
28  }
29
30  //---------------------------------------------------------
31  void loop() {
32
33    // set FIRE_OUT HIGH only if FIRE_IN pressed
34    digitalWrite(FIRE_OUT, HIGH & digitalRead(FIRE_IN));
35
36    // set Ready LED HIGH as duty_on indicator
37    digitalWrite(READY_LED, HIGH);
38
39    // read pot and wait for duty_on time
40    delay(map(analogRead(DUTY_ON),0,1023,scale_l,scale_h));
41
42    // set FIRE_OUT LOW (always safe to do)
43    digitalWrite(FIRE_OUT, LOW);
44
45    // set Ready LED LOW as duty_off indicator
46    digitalWrite(READY_LED, LOW);
```

```
47
48     // read pot and wait for duty_off time
49     delay(map(analogRead(DUTY_OFF),0,1023,scale_l,scale_h);
50   }
```

Lines 1–8: Setup Aliases

```
1    // Define pin names
2    #define FIRE_OUT 8
3    #define FIRE_IN 9
4    #define DUTY_ON 0
5    #define DUTY_OFF 1
6    #define READY_LED 10
7    #define scale_l 100
8    #define scale_h 2000
```

One of the ways that Arduino sketches allow substitution of
a human-readable name for pin numbers and other static values
is the #define preprocessor directive. #define works by directing
the environment to perform a text substitution at compile time.
It takes the first string provided (the identifier) and replaces it
everywhere in the code with the second string (the replacement).
This makes for a very lightweight mechanism for things like
naming pins, but it means that you have to be careful not to use
the identifier string any place you don't want it swapped, such as
inside another string.

For example, if I provide a directive of

```
#define FIRE 10
```

then any place in the code that the string FIRE appears will
be changed to 10. So I better not name something FIRE_OUT or
it will become 10_OUT'. Use unique strings if you use #define
directives.

We could have used variables (vars) as names and assigned
the pin numbers as values, but I retain habits from a time when
memory was precious and I try not to use it when I have a good
alternative. On lines 7 and 8, I'm providing the new ranges for
modifying the analog inputs. These should be vars if I want to
be able to change them with code. Since I'm using static values
changed at compile, I can use #define.

This block finds locations in the code where strings like FIRE_OUT and DUTY_OFF are used and replaces those strings with numbers. Those numbers are the pin numbers we want used when we refer to the strings.

Lines 11–28: Initial Setup Function

```
11  void setup() {
12
13    // Initialize pins
14    pinMode(FIRE_OUT, OUTPUT);
15    pinMode(FIRE_IN, INPUT);
16    pinMode(READY_LED, OUTPUT);
17
18    // set FIRE_OUT LOW (a good safety step)
19    digitalWrite(FIRE_OUT, LOW);
20
21    // 4 blinks of Ready LED as Power On Self-Test (POST)
22    for (int i = 0; i < 4; i++){
23    digitalWrite(READY_LED, HIGH);  // turn the LED on
24    delay(500);              // wait for a second
25    digitalWrite(READY_LED, LOW);  // turn the LED off
26    delay(500);              // wait for a second
27    }
28  }
```

Next we'll do the setup( ) function. This function runs once before we enter the main loop. It's the place we do initializations and boot activities.

Lines 14–16 declare the three pins we'll be using as either an INPUT or an OUTPUT. We're using the pin names we declared in the #define directives; these will be swapped out for pin numbers at compile.

Line 19 is a little bit of extra insurance (always good to have in systems that have safety concerns). This line explicitly sets the FIRE_OUT pin, which connects to the MOSFET, LOW. We do this at the very first opportunity to avoid a floating value on the pin. We added insurance in the hardware against this concern with a pulldown resistor. This line in the code may be technically redundant, but safety concerns should be reflected in both hardware and software.

Lines 22–27 blink the ready LED four times over the course of four seconds (half a second on, half a second off) by setting the READY_LED pin HIGH, waiting, then setting it LOW and waiting. This is useful as a means to let us know, on power-up, that the microcontroller is healthy and executing our code.

Lines 31–50: Main Loop Function

```
31  void loop() {
32
33    // set FIRE_OUT HIGH only if FIRE_IN pressed
34    digitalWrite(FIRE_OUT, HIGH & digitalRead(FIRE_IN));
35
36    // set Ready LED HIGH as duty_on indicator
37    digitalWrite(READY_LED, HIGH);
38
39    // read pot and wait for duty_on time
40    delay(map(analogRead(DUTY_ON),0,1023,scale_l,scale_h));
41
42    // set FIRE_OUT LOW (always safe to do)
43    digitalWrite(FIRE_OUT, LOW);
44
45    // set Ready LED LOW as duty_off indicator
46    digitalWrite(READY_LED, LOW);
47
48    // read pot and wait for duty_off time
49    delay(map(analogRead(DUTY_OFF),0,1023,scale_l,scale_h));
50  }
```

This function, called loop( ), is run continuously after setup() has completed. It is the heart of an Arduino program. We're actually doing very little in it, so it should run quickly and without problems. Let's go through it line by line.

```
34    digitalWrite(FIRE_OUT, HIGH & digitalRead(FIRE_IN));
```

This statement performs a *digitalWrite*, meaning that it sets the state of a pin. We have designated FIRE_OUT (pin eight; see line 2) as the pin to write to. Setting that pin HIGH will switch our MOSFET on and open the solenoid valve. We tell it to set it HIGH, but add a condition.

The & symbol means that we want to do a Boolean AND operation. We often use what's called a *truth table* to describe how Boolean operations work. This is because, in our case, TRUE is the same as HIGH, which is the same as 1. FALSE is the same as LOW, which is the same as 0. The truth table for AND looks like this:

1 & 1 = 1	HIGH & HIGH = HIGH	TRUE & TRUE = TRUE
0 & 1 = 0	LOW & HIGH = LOW	FALSE & TRUE = FALSE
1 & 0 = 0	HIGH & LOW = LOW	TRUE & FALSE = FALSE
0 & 0 = 0	LOW & LOW = LOW	FALSE & FALSE = FALSE

The only combination of AND that will result in 1 (aka HIGH, or TRUE) is when both items are equivalent to 1.

There are other Boolean operators that will provide different results. I would argue that understanding Boolean logic is the single most important thing you can learn as a programmer (though certainly not the only thing). If you're lucky, someday you might get to use the mighty XNOR!

In our case, we set the FIRE_OUT pin to the result of HIGH and the state of the FIRE_IN pin. So if FIRE_IN is HIGH the result is HIGH (boosh), but if the FIRE_IN is LOW (meaning we're not pushing the fire button) then the result of the AND is LOW (no boosh).

Since this line is the only place in our code where we might set FIRE_OUT to HIGH, opening the solenoid, this is the only time we need to check if the fire button is being pressed. It's always okay to set FIRE_OUT to LOW; that just means we are closing the valve, which is its default state.

```
37    digitalWrite(READY_LED, HIGH);
```

It's handy to have a way to visualize the duty cycle without having to fire the boosh. We'll use the ready LED as a visual indicator. This line does the same thing as line 32, but doesn't check the FIRE_IN switch status. We want the LED to trigger with each cycle so that we can adjust the duty pots to get the pattern we want before firing.

```
40    delay(map(analogRead(DUTY_ON),0,1023,scale_l,scale_h));
```

Having set the LED, and possibly the FIRE_OUT pin, to HIGH, we need to have them remain so for the length of time we desire. This time is derived from the DUTY_ON analog input.

The values returned from analogRead range from 0–1023. We want to use the value returned as the number of milliseconds to delay (hold the valve open). But that's pretty limiting. If we want our on-time (or off-time) to be between 0 and 1.5 seconds or 0.1 and 10 seconds, we can't do it if the minimum value returned is 0 and the maximum is 1023 (which results in 1.023 seconds).

Luckily, Arduino provides the map( ) function. This function takes five arguments. The first is the value being measured. The second is fromLow, which is the lower end of the range of values being measured, in our case 0. The third is fromHigh, which is the upper end of values being measured, in our case 1023. The function then maps this range onto a new range with the fourth and fifth arguments, toLow and toHigh.

We'll use map( ) to change 0–1023 into the values we provide as scale_l and scale_h in the #define section. This gives us an easy way to change the range. It could be 0–500 milliseconds, 100–2000 milliseconds, or any other range we desire.

Lines 43, 46, and 49 do the same thing we just described, but with the FIRE_OUT and ready LED being set to LOW. We don't have to check the FIRE_IN on line 43 since we are always okay with shutting the solenoid valve.

That's all there is to the code. The main loop executes continuously after boot, so it's constantly generating the duty cycle pattern. All you have to do is press the fire and safety buttons while in automatic fire mode to fire your boosh with that pattern. If you want to use the boosh in "normal" mode, you can use the auto/manual switch to avoid the duty cycle altogether.

You can upload this code to the Arduino (or equivalent) at this time.

Testing

Testing the controller is straightforward. You don't have to have the components all assembled into a case to do it, though you may want to do some quick testing again once you have.

Wiring Test

We can quickly test all the soldering on the protoboard without having it powered up or attached to the Arduino. Using a multimeter in ohm-meter mode, we can test between sets of points to verify if the soldered joints in between are good. We already tested the screw terminals and headers early on when we assembled the ProtoScrewShield, so we don't need to test them again (though it won't hurt).

Many multimeters have a continuity tester built in, which is really just an ohm meter that distinguishes between infinite resistance (open circuit) and no resistance (closed circuit). We'll be testing through components like resistors, so we'll also need to use the ohm-meter function. The expected resistances between test points are listed next. CONT indicates continuity, in other words, no resistance, meaning the path is connected.

If the readings between test points aren't as noted, check out the diagrams (see following table for guidance) to see the specific intermediate points to test at to troubleshoot. If you find a problem, resolder the junction and test again.

STEP	TEST POINT 1	TEST POINT 2	METER	FIGURE
1	Solenoid LED+	V_{IN}	CONT	Figure 11-16
2*	MOSFET drain (tab)	V_{IN}	Diode	Figure 11-17
3	MOSFET drain (tab)	Solenoid LED–	470 ohm	Figure 11-18
4	Ready LED –	Ground	330 ohm	Figure 11-19
5	Auto fire return	Ground	1000 ohm	Figure 11-20
6	MOSFET supply	Ground	CONT	Figure 11-21
7	MOSFET gate	Ground	1150 ohm	Figure 11-22
8	Safety return	Manual fire return	CONT	Figure 11-23

* If your multimeter doesn't have a diode test range, you can test between the MOSFET drain (tab) and the positive lead (no line) of the diode for continuity and then from the diode's negative lead (line) to V_{IN}.

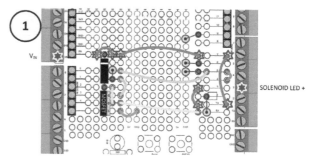

FIGURE 11-16: 1. Solenoid LED+ to V_{IN}

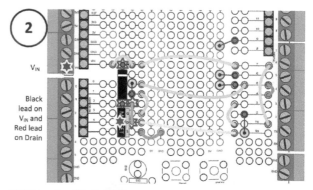

FIGURE 11-17: 2. MOSFET drain to V_{IN}

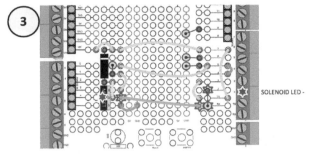

FIGURE 11-18: 3. MOSFET drain to solenoid LED–

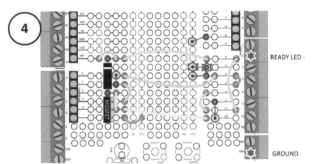

FIGURE 11-19: 4. Ready LED– to ground

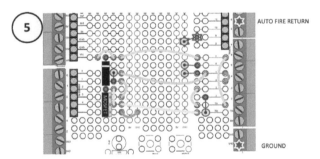

FIGURE 11-20: 5. Auto fire return to ground

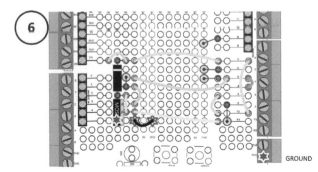

FIGURE 11-21: 6. MOSFET supply to ground

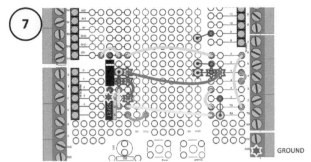

FIGURE 11-22: 7. MOSFET gate to ground

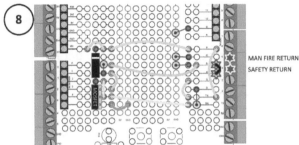

FIGURE 11-23: 8. Safety return to manual fire return

Once you have confidence that there are no shorts or bad solder joints and everything is wired appropriately, you can move to the next set of tests.

Smoke Test

Traditionally, when building electronic projects (at least in the maker circles I've grown up in), the first test is to see if any of the magic smoke escapes when the circuit is powered up. This is predicated on the idea that electronics operate using compressed magic smoke and, should the smoke get out, they won't work anymore. If you don't care for the magic smoke worldview, just know that when you power up the circuit and you smell or see smoke, you know you have a serious problem.

5V Power Test

Attach the protoboard to the Arduino (or equivalent). Do not connect any outboard components yet. Connect the Arduino to an appropriate 5V power source or USB connection. Sniff around the board and look for any smoke.

Safety Note

MOSFETs have the potential to get hot enough to cause serious burns. Always be extremely cautious when checking to see how hot a MOSFET is. Never just reach down and touch a MOSFET that has been powered up. Hold something small (and not flammable or conductive) against the MOSFET for a moment, then feel it for retained heat. A pencil eraser is useful for this. If you must check it by hand, always do so with an initial quick brushing motion first.

COMPONENT CONNECTION	SCREW TERMINAL LABEL
Left Side of ProtoScrewShield	
Fire send	5V
Duty pot +	5V
Duty pot –	Ground
Duty on	0 (Analog in)
Duty off	1 (Analog in)
Right Side of ProtoScrewShield	
Ready LED+	10
Auto fire return	9
Safety send	8
12V+ input	7
Ready LED–	6
Manual fire return	5
Safety return	4
Solenoid LED+	3
Solenoid LED–	2
12V+ to solenoid	TX
Ground to solenoid	RX
12V– (ground) input	Ground

12V Power Test

Disconnect the 5V or USB power source. Connect all the outboard components to the appropriate screw terminals.

Connect the solenoid connector to a solenoid. If you have a solenoid with the connector attached and not plumbed into a system yet, you can connect it directly without using an extension cable. Otherwise, run an extension cable to a plumbed solenoid. Connect the 12V power cable, but do not plug it into the 12V source.

Depress the emergency cutoff switch or use the key switch to make sure power won't flow to the board, and connect the power cable to the power source. Make sure the source is active, and unlock the cutoff or turn the key to provide power to the system.

If the answer is "No" to any of the following questions, turn off the power immediately.

A. Does the ready LED start blinking immediately?

B. Is the device smoke- and smell-free?

C. After the boot-up blinks, does the ready LED change its blink pattern in response to changes of the duty on and duty off LEDs?

D. With the auto/manual fire select switch on manual, does pressing and releasing the fire switch cause the solenoid LED to light and the solenoid to engage and disengage?

E. With the auto/manual fire select switch on automatic, does pressing the fire switch **without** the safety pressed do nothing?

F. With the auto/manual fire select switch on automatic, does pressing the fire switch **with** the safety pressed cause the solenoid LED to light and the solenoid to engage in time with the ready LED?

G. Is the MOSFET cool enough to not burn you? (Remember the safety warning about testing this!)

If the answer was "Yes" to all the tests above, the system passes and is ready to go! If the answer is "No" to any of them, troubleshooting is in order. Here are some troubleshooting tips.

There are other possible problems that are fairly uncommon; try using the multimeter to test the wiring with the power off. If necessary, de-solder components using either de-soldering braid or a suction rig. I'm old-school and tend to use braid.

CONDITION FAILING	POSSIBLE PROBLEM
Any	**Wiring.** Recheck the board wiring and verify the solder connections on all the outboard components.
A, C, E, F	**Code problem.** Check that the software matches the code in the book; fix and upload if necessary.
B, G	**Polarity problem or a short.** Verify that nothing positive is connecting to anything negative. Check the polarity of the diode. Use a continuity tester, with everything connected, but power turned off, to be sure that 5V and ground are not connected somehow. Do the same for 12V+ IN and ground.
C	**Pot wiring.** Check that the potentiometers are getting 5V and ground; check the voltage at the center tap to be sure they are working as voltage dividers.
D	**Outboard wiring.** Check the LED, fire switch, and solenoid connector soldering.
E, F	**Outboard wiring.** Check the safety and auto/manual fire select switch wiring.

To de-solder with braid, hold a clean section of braid against the point to de-solder—keep your hands a few inches away! Press a hot soldering iron tip just slightly up the braid from the desired de-soldering point; this is what will provide the pressure of the braid against the solder you want removed. The goal is to have the iron heat the braid hot enough to melt the solder. The braid will wick the solder, once liquid, up into its weave. The braid gets hot, so be careful.

Other people swear by suction tips that will vacuum up the melted solder, manually or with power. Check out videos online to get a good idea of how to de-solder effectively.

Casing the Project

As noted earlier, this chapter isn't going to specify how to case this project. Chapter 9 goes into detail about how to prepare a case and mount components. For this project, there are too many options or enhancements you might want to make differently. I will, however, note some important considerations in your case (and component) choices.

Sliding versus Rotary Potentiometers

You may have chosen rotary or sliding potentiometers. From an electrical standpoint, these are identical. Rotary pots have the advantage of being extremely easy to mount; a single small hole in the case suffices. Sliding pots have a visual appeal that is hard to deny for applications like this, because they provide a powerful tactile and visual feedback of their state that is hard to get with a knob on a rotary pot. (See Figure 11-24.)

FIGURE 11-24: Rotary versus sliding pots

Sliding pots are much more difficult to mount, usually requiring a slot and two screw holes. If you are using a metal case and you decide to mount sliding pots, invest in a sheet metal nibbler hand tool. This tool will cut squares or slots or curved patterns in light sheet metal. There are electric and air-powered versions that are wonderful to use, but a cheap hand nibbler is always a great tool to have around.

Sliding pots also tend to be a magnet for dirt, gunk, and liquids. A rotary pot is, for our purposes, sealed, whereas a sliding pot leaves its resistance track open to the world. Dirty sliding pots are all too common, especially when used in outdoor situations. If you do end up using sliding pots, try to seal the slot by putting heavy tape (electrical or duct) on both sides of the slot and then cut a slit down the middle for stem of the pot to move through. This will reduce the amount of stuff that can get down into the tracks.

Standoffs and Ground

In Chapter 9, it was sufficient to make sure that none of the wires touched the case. We even added a layer of duct tape to insulate the lid from any possibility of an over-excited boosher squeezing the case so hard that contacts shorted against it. Now that we are using a printed circuit board, we have to go to greater lengths to make sure that the contacts on the board do not short out against the case.

You may be way ahead of me here and have decided to use a 3D printed custom case or a plastic project box, or to seal the whole project in hot glue (which is interesting, but difficult to troubleshoot). Using a metal case means using standoffs to keep the board's contacts from shorting out. We use these in the mounting holes of the PC board (in this case, the Arduino) as a means to hold it in place some distance above the case.

Standoffs come in a wide variety of forms. They can be metal, nylon, or home-made. (See Figure 11-25.)

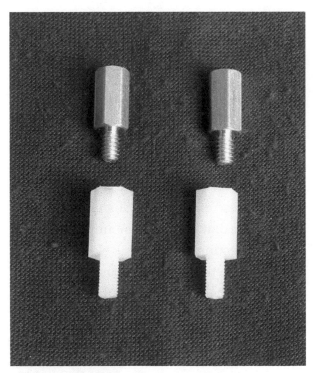

FIGURE 11-25: Metal and nylon standoffs

The most commonly used standoff is the metal hex spacer. However, in this project, it's worth thinking through the ramifications of using one of these.

The mounting holes on most printed circuit boards have a *pad* on them. This is a copper ring that is often electrically connected to some part of the circuit. In most cases,

the mounting holes are connected to ground (V_{ss}). If used with metal standoffs, this electrically grounds the circuit board to the project case. This is usually a desirable feature. In our case, it bears consideration.

The MOSFET we're using is in a *low-side switch* configuration, meaning it provides ground to the load, sinking current. This is the easiest method for using N-channel MOSFETs. Using N-channel MOSFETs for a *high-side drive* (sourcing the positive voltage to the load) requires a more complicated gate driver. Alternately you could use a P-channel MOSFET, but they have their own issues. Rather than insult any electrical engineers with my attempt at explaining why, search the Internet for "N-channel MOSFET high-side switch" and you'll find a wealth of explanations. N-channel low-side MOSFET circuits are common and efficient, so that's why I chose one for this controller.

In any event, ground in our circuit isn't completely "common." If anything in the drain path of the MOSFET is shorted to ground, the solenoid will have an active circuit and open (venting gas accidentally, which is one of our big things to avoid). As we'll talk about more in the "Heat" section below, N-channel MOSFETs have their mounting tab (in the T-220 package we're using) internally attached to drain (the lead connected to the solenoid). *This means that there is a big metal tab sticking up from the shield that must **not** be inadvertently shorted to ground.*

This makes it undesirable to connect the ground plane of the Arduino (and the shield mated to it) to the case. We don't want the case to be a big switch. All of this is the reason that I will recommend using nylon standoffs instead of metal standoffs to mount the Arduino to the case.

Heat

The load and usage of a single solenoid should easily be within the capabilities of the N-channel MOSFETs I have specified. A ¾" 12V DC solenoid typically draws around 2 amps; usually less. We also aren't in a position where it will be in continuous use for more than a few seconds. Nevertheless, power MOSFETS switching amps create heat. If we were trying to drive a heavier load (greater current), we would need to start designing solutions for heat management. Though this circuit shouldn't need heat management, it's worth understanding what to do if you want to include it, since you may add more or bigger solenoids in the future.

Since heat is inevitable (thanks, thermodynamics!) our task is to dissipate it. The standard practice, which the T-220 case is optimized to support, is to add a heatsink. This is a piece of metal, frequently with fins, that attaches to the back of the T-220 case and provides a greater surface area for the heat to disperse from. Heatsinks are most efficient when a thermal paste is used to improve the contact between the case and the sink. (See Figure 11-26.)

The grounding problem discussed in the section above is relevant to the heatsink discussion, as well. Mounting a heatsink directly to the case causes it to be electrically bonded; in our case, to the MOSFET drain. Luckily,

FIGURE 11-26: A T-220 heatsink

this is a problem with a known solution. Heatsink insulator mounting kits are easy to acquire. They consist of an insulator sheet to go between the heatsink and case (these used to be mica but are more commonly silicone these days) and an insulating shoulder washer that keeps the mounting screw from connecting the heatsink electrically to the case. It's still advisable to use thermal paste on both sides of the insulator sheet to improve the heat transfer. (See Figure 11-27.)

FIGURE 11-27: Insulating heatsink mounting kit

Ultimately, even a heatsink won't help if there is no way to move the heat from the sink or case somewhere else. Air is the most common transfer mechanism for doing this. However, air standing still is not that useful—the air must be moving so that convection will work effectively for us. Cooling slots or a grill in the side of the case is one way to add some air flow, but a fan is the most effective.

Computer fans come in all sizes, even tiny 25 mm square ones for 12V or 5V use. These draw milliamps of power, so are easy to use with small circuits. Wiring one of these to the 12V input and mounting it on the case (with the appropriate sized hole) will add a great deal of heat transfer. It will be much more efficient if you add a grill on the other side of the case and mount the fan so that it blows outward, which will draw air in through the grill, across the heatsink and the MOSFET, or both, and out the fan.

Weather

Unless you go to great lengths, many of which are counter-productive from a heat-management standpoint, the controller case is not waterproof. You neither want to expose it to, nor operate it in, the rain. However, there is another, more challenging, circumstance than rain to consider—the infamous dew point.

Booshing is frequently a nighttime endeavor. The flames are more dramatic at night in the dark and, at festivals, this is when the party is most aggressively underway. Yet nighttime is when the dew point stalks the unwary boosher.

The dew point is water's saturation temperature in air. Varying with humidity, above this temperature, water will evaporate, below this temperature water will condense. The higher the humidity, the higher the dew point will be for a given temperature.

It is startling when, on a beautiful night with the temperature dropping, you reach the dew point. Your kit, and everything else, becomes soaked, inside and out. This is not the best situation for electronic components, especially if the humidity is high enough that the water comes pouring off, and out of, the case.

The dew is purified water, and a pretty good *dielectric* (doesn't conduct electricity). So for the most part, it won't cause a short circuit by itself. The risk is that the accumulated dew will pick up impurities from the board or some other source that will give it electrolytic properties and allow it to conduct electricity.

So what can you do about this? Here are a few options, all of which have challenges.

1. **Stop using the device.** Take a break, unplug from the power source, and wait until the water has all condensed out of the atmosphere. Give the controller a chance to shed all its accumulated water. Don't leave it in a position where the circuit boards are horizontal. Downside: Having to stop when you're having fun.

2. **Keep using the device.** If the MOSFET is generating heat in the case, it may keep the internal temp above the dew

point. It may be worth having a mechanism to disable the fan (if you have one) for this circumstance. If the components are all clean, the likelihood of the dew conducting electricity is low. Downside: If too much water builds up in the wrong place and it manages to conduct electricity, the possibility of a short circuit increases.

3. **Seal the case.** This is harder than it sounds and generally in conflict with heat-management goals. There are special connectors for sealed connections and you can seal around components with silicone or hot glue (thermoplastic). Downside: It's expensive, and it can be difficult to be effective and manage heat.

4. **Coat the at-risk components**. Products known as *conformal coatings* and *humi-seals* can be sprayed on circuit boards to protect them from water. Experimenters have even tried dipping boards in urethane and varnish. Outboard components such as switches can be covered in silicone or thermoplastic. *Potting* boards by putting them in a tub and pouring epoxy over them until they are covered is a traditional technique that works (though once again, heat management can become an issue if heatsinks get covered). Downside: It's a great big hassle and mess, and makes troubleshooting or repairing the boards a nightmare, if not impossible.

All of these are a bummer to one degree or another. I confess to doing the second more often than I'd like to admit, though the first is my preference. You could add an internal humidity detector and heater (though I'm not sure what I'd advise you to safely use as one) or you could come up with other solutions. All in all, it's a personal choice. I hope understanding the situation will help you make a good one.

Enhancements

Now that we have a microcontroller as part of the firing circuit, there are an almost unlimited number of new possibilities. I'll break some suggestions into hardware and software categories (though the hardware and software aspects rapidly intermingle).

Hardware

In Chapter 10, "Pilot System," we incorporated an automatic pilot igniter. By plumbing a solenoid valve into the pilot feed line, we could eliminate the always on pilot, and only ignite it just prior to lighting the main effect.

While you always want visual indication that the pilot is lit (due to the flame effect group we're working in), you could also add a flame detection sensor for the pilot and automate the entire sequence. With a single button push, you could:

1. Open the pilot feed valve.

2. Ignite the pilot.

3. Verify pilot ignition.

4. Open the main effect valve (perhaps with pattern fire).

Adding multiple solenoids and the MOS-FETs to control them is relatively straightforward. Be careful of the inadvertent grounding issues we discussed above and be careful of heat management. If you control a pilot solenoid with a MOSFET, it is likely to be open longer than the main effect valve and get significantly hotter. Heatsinks become much more important.

As you add additional mechanisms for the microcontroller to manage gas paths, think long and hard about how you will design safety systems, such as the dead man's switch we discussed earlier. Always provide the ability to immediately disengage all valves in an emergency.

If you looked closely at the PCB included earlier, you'll notice a location for a chip labeled *micro_c*. This is because the Arduino is, in many ways, overkill for this effort. In many instances, single-chip microcontrollers such as PICs, PICAXE, BasicATOM, and others can do the job just as well. This allows smaller custom PCBs to be designed. In many cases, eight-pin microcontrollers are often cheaper than single-purpose 74-series TTL chips (used for AND gates, etc.).

Software

Firing the boosh based on a duty cycle is very cool. But patterns are software's bread and butter, so many other things are possible. The first enhancement I've made to my control software is to add single-shot, tri-shot, and auto-fire options.

By adding a switch (rotary encoder, additional pot, etc.) to communicate mode selection to the controller, I can then provide alternate functions for the code to call. Single-shot uses the duty cycle length, but only fires one pulse when the fire switch is engaged. Tri-shot fires three duty cycle–defined bursts, and auto-fire acts like the project software described above, firing for as long as the fire button is pushed.

The duty cycle is also subject to modifications in code. You could use it as a starting point to fire successively shorter or longer bursts. You could fire random patterns or have the duty cycle derive from some analog sensor that detects noise peaks. Imagine an installation that generated boosh sizes in response to audience noise. A boosh-based applause meter!

This project is really the gateway to untapped wealth of interesting flame effects. I hope it has spurred your interest and that you'll go build something safe and fantastic!

12

The Digital Step Valve

THIS FINAL PROJECT IS an example of a different direction you can take with flame effects. I've mentioned in other chapters that, in most flame effect hobby communities, there seems to always be an arms race for the biggest boosh. But there are many other directions you can go, as well. I've seen micro-booshes on RC cars, fiery fingertips on robot hands, and interactive flame art aware of the actions of the people around it.

The *digital step valve* (DSV) is an experiment in providing more values than just on and off for a flame effect. There are industrial solutions, usually referred to as proportional valves, that offer variable output, but they are typically expensive, relatively slow, and frequently require expensive specialized controllers. The DSV takes a different route. Rather than attempting to create analog control over the valve, the DSV uses multiple valves to create a series of digitally addressable output volume "steps."

The DSV uses a *manifold*, a pipe that has multiple outlets, to split the incoming vapor into four paths. The paths will use needle valves to allow a different amount of propane to pass through each channel. We can control which path is active by use of solenoid valves.

The first path has only a needle valve, which is the "pilot" path and is always on at a minimal level. The last path has only a solenoid valve and is the "full-on" path that represents our maximum output. The two middle paths have both a needle valve and a solenoid valve, with each needle valve set to a successively higher output.

By changing which, if any, solenoids are open, we can change the output pressure of the final manifold between the four levels very quickly. We can feed this output into a burner, a variation of the flambeau, so that we can make a flame curtain jump between levels at higher speeds than most proportional valves can achieve.

There are many ways to incorporate a multistep flame curtain into a flame effect art installation. Formed into a ring, it makes a great alternative pilot for a boosh, while providing an entrancing flame effect in between big fireballs. Controlled by software, the DSV can jump a flame in time to music. Your imagination will come up with many more ideas as you consider the possibilities.

The DSV is expandable. We'll start with four stages, but it's straightforward to add more steps with additional manifolds. Since this isn't an attempt to chase high volume, we can use lower cost, smaller solenoid valves. The only unique skill set required is something we learned in the boosh chapter—flaring copper tubing—so there are no special challenges. The DSV is a mid-level, fun project that will get you thinking about new ways to produce safe fire art, I'm looking forward to seeing what you come up with!

Parts

In many ways this project is a celebration of fittings. There are innumerable ways to configure the set and ordering of the parts involved in this project. I have done my best to present the smallest viable parts count I can come up with. (See Figure 12-1.)

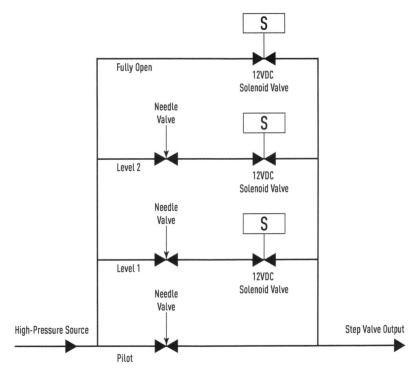

FIGURE 12-1: Digital step valve schematic

Step Valve Parts:

REF	ITEM	QTY
F1	Manifold ½″ FIP × 6	2
F2	Plug ½″ MIP	2
F3	Brass adapter ½″ MIP × ⅜″ MFL	7
F4	Brass adapter ¼″ MIP × ⅜″ MFL	6
F5	Brass bushing ¼″ FIP × ½″ MIP	1
F6	Brass flare nut ⅜″ FFL	4
F7	Brass swivel flare union ⅜″ FFL × ⅜″ FFL	4
F8	Pipe nipple ½″	1
F9	Brass adapter ¼″ MIP × ½″ MIP	1
P1	Copper tube ⅜″ OD × 12″ (will be cut into 2 pieces)	1
V2	Needle valve ¼″ MIP × ¼″ FIP	3
V3	12V DC solenoid valve ¼″ FIP × ¼″ FIP	3
V4	Ball valve ½″ FIP × ⅜″ MFL	1

Optional Controller Parts:

REF	ITEM	QTY
C1	Male four-pin trailer connector	2
C2	Female four-pin trailer connector	1
C3	74x139 dual 2–4 decoder	1
C4	74x04 hex inverter	1
C5	N-channel power MOSFET (see Chapter 11, "Arduino Control")	3
C6	7805 voltage regulator	1
C7	SPST switch (non-momentary contact)	1
C8	LED	4
C9	100 ohm resistor	4
C10	1K ohm resistor	2
C11	1N4001 diode	4
C12	Case, standoffs, protoboard	1
C13	1" split-ring hangers with ⅜" x ⅝" bolts	2

The diagram in Figure 12-2 represents the parts and their connections.

There is one new tool that will aid in construction of the DSV. You may have used or own some already—digital calipers. I'm going to short-circuit a long debate about digital versus analog calipers. If you want to wade into that debate, you'll find numerous opportunities online. Digital calipers, even the amazingly cheap ones, are very useful. You won't regret having more than one pair around if you also feel compelled to buy expensive calipers that you're willing to be seen in public with. (See Figure 12-3.)

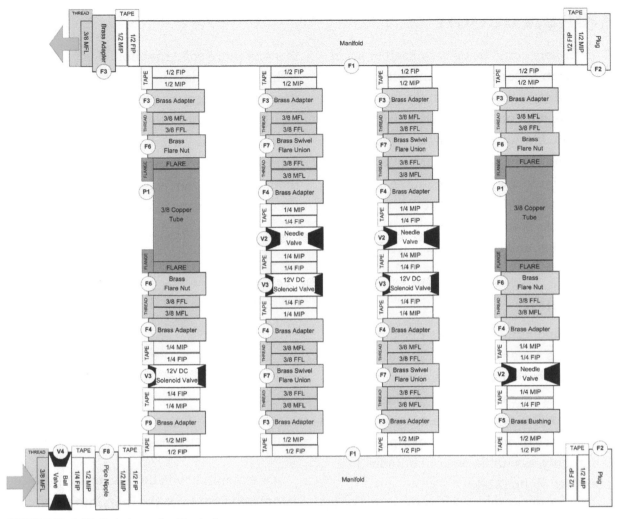

FIGURE 12-2: Digital step valve block diagram

FIGURE 12-3: Digital calipers

Construction

This project could be used by itself or as an expansion to some other flame effect project. You could use the DSV with the flambeau we built, the pilot on your boosh, or with a more elaborate burner you construct. There are many choices, all of which potentially modify what to build. I will document the construction of the DSV itself and provide a quick pass across some of the other parts, such as a controller and burner, that you might want to build or get ideas from.

Plumbing

The component that makes this project feasible is a manifold. A *manifold* essentially splits or combines some number of pathways. There are many ways to fabricate a manifold. You can machine one out of a solid block, you can weld one out of pipe, or most commonly, assemble one out of individual tees. (See Figure 12-4.)

However, as we'll discuss throughout this chapter, tolerances need to be fairly tight for this project to work. This gives an advantage

FIGURE 12-4: **Various gas-rated manifold approaches**

to a one-piece solid manifold design. So rather than fabricating or assembling manifolds for this project, I recommend buying one. Luckily, a six-outlet solid manifold has become generally available on the market from Pro-Flex®. It is available in ½" and ¾" models. The ½" model is the manifold (F1) used in this project. (See Figure 12-5.)

FIGURE 12-5: **Pro-Flex® cast iron CSST manifold**

One manifold will be used to split the incoming vapor from the inlet into four paths. The other manifold will recombine these four paths and send them to the outlet. We'll have an extra hole in each manifold that could be used for expansion or even pressure gauges.

The Send Manifold

We'll start by assembling the send manifold and its associated parts. Tape a ½" plug (F2) and thread it tightly into one end of the manifold. Tape both ends of a ½" pipe nipple (F8) and thread one end tightly into the other end of the manifold. Tighten a ½" FIP × ⅜" MFL ball valve (V4) onto the remaining threaded end of the pipe nipple.

Tape both threads of a ½" MIP × ¼" MIP adapter (F9) and thread one end very tightly into one of the remaining manifold outlets. I will fill these outlets starting from the ball valve end, but this is arbitrary. Tighten a ¼" FIP × ¼" FIP 12V DC solenoid (V3) onto the taped ¼" MIP end of the adapter. It's aesthetics, but I worked to get the coil of the solenoid (the black cylinder) pointing upward. If, in testing, you determine that these joints leak, you'll have to tear most of the device down to retighten, so it's worth making an effort to carefully tape and completely tighten at this stage. Be careful to note the flow arrow on the solenoid and position it appropriately. (See Figure 12-6.)

FIGURE 12-6: The solenoid on path one

Tape the threaded (not flared) end of a ½" MIP × ⅜" MFL adapter (F3) and thread it tightly into one of the central manifold outlets. Repeat this for the other central outlet. It is important to note that the two ½" MIP × ⅜" MFL central adapters need to be threaded to the same depth. Different manufacturers

have different dimensions for the total height of their adapters, so it's important, if not critical, to use matching adapters. Use your calipers to measure between the back of the manifold and the orifice of the adapter. Continue to tighten in small increments until you have them both protruding the same height above the manifold body.

For the final outlet, tape the external threads of a ¼" FIP × ½" MIP brass bushing (F5) and thread it tightly into the outlet. Tape the ¼" MIP threads (which are on the inlet) of the needle valve (V2) and thread it tightly into the bushing. These are all the rigid components on the send manifold. (See Figure 12-7.)

As noted, the way that the solenoid coils are oriented is primarily an aesthetic choice. However, two practical factors come into play. The first is that a solenoid and a needle valve cannot be side by side and allow the needle valve handle to fully rotate, which is a requirement. The second factor is that, since the needle valve handles are a significant control surface, it is reasonably important

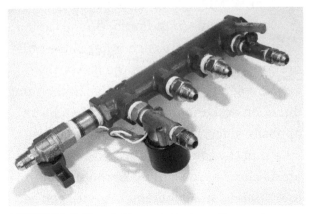

FIGURE 12-7: The rigid components on the send manifold

to position them in a manner that makes it easy to adjust them. For these reasons, I've built my DSV so that the coils point out to one side of the device and all the needle valve handles point out the other side.

If you're like me, you'll look at this design and think that there must be a way to do it with fewer parts. The biggest challenge in doing so is that the solenoid coils are too big to allow the solenoid to rotate to thread it into the other parts without hitting the components of the next path over. The design in this book gets around this by only having one solenoid connected directly to the manifold with pipe fittings. The other two are connected with swivel fittings that allow the joint to be tightened without rotating the solenoid.

The Central Assemblies

Since both of the two central paths combine a solenoid and needle valve, we will build two identical assemblies. There are four parts in each unit. In order, they are a ¼" MIP × ⅜" MFL adapter (F4), a 12V DC ¼" solenoid (V3), a ¼" MIP × ¼" FIP needle valve (V2), and another ¼" MIP × ⅜" MFL adapter (F4).

Tape all male (non-flared) threads and tighten the parts into each other in the order described. Be sure to align the flow arrows on the solenoid and needle valve in the same direction and orient the coil and handle on opposite sides. Use your calipers to make sure that the total lengths of the assemblies are as close as you can make them. Continue to tighten in small increments until you get them within a few thousands of an inch. (See Figure 12-8.)

FIGURE 12-8: The central assembly

The Receive Manifold

We'll start the construction of the receive manifold by taping and tightly threading a ½" plug (F2) into one of the ends. Tape the threads of a ½" MIP × ⅜" MFL adapter (F3) and tightly thread it into the other end.

All four of the inlets on the manifold will get the same ½" MIP × ⅜" MFL (F3) adapter taped and tightly threaded into them. You'll notice in Figure 12-9 that I was unable to get four matching adapters. The important thing is that the two center adapters are at the same height. Use your calipers to tighten them in small increments until they are at the same level.

Mount the Central Assemblies

When it comes to mounting the central assemblies, we're once again up against

FIGURE 12-9: The receive manifold

subtle manufacturer differences in dimensions. We need both central assemblies to end up the same length, so if the ⅜″ swivel flare adapters are of different lengths, we're in trouble. Shop carefully so you can get four that match or two pairs that match.

Using the ⅜″ swivel flare adapters (F7), connect the central assemblies to the flare fittings on the two central paths of the send manifold. Tighten the swivel flare nuts so that the central assemblies are mounted firmly. Tighten a ⅜″ swivel flare adapter to the end of each assembly. (See Figure 12-10.)

Once the assemblies are mounted, pull the flare nuts on the ends as far as possible and hold a ruler across one of them. Ideally the other nut will be flush against the ruler's straight edge. If not, tighten the various flare fittings in the stack to get the two nuts as close as possible.

Fabricate the Outside Connecting Tubes

We'll use the flaring techniques described in Chapter 9, "The Boosh," to construct tubes that mount the outside paths to the receiving manifold, so don't hesitate to look over that chapter again for more details. The length of the tubes is extremely important. To determine the length, we need to attach the central assemblies to the receive manifold. Tighten the flare fittings firmly on both assemblies and orient both manifolds so that they are on the same plane.

Use your calipers to measure the distance between the orifice openings on both outside paths. (See Figure 12-11.)

FIGURE 12-10: The mounted central assemblies

FIGURE 12-11: Measure the outside tubes.

Your distances will depend on the specific fittings you're using. In my case, the distances were 4.635″ and 5.236″. We have to add ¼″ to each measurement to provide a ⅛″ flare on each end. This brought my measurements to 5.486″ and 4.885″. Set your caliper to one of these lengths (or as close as you can get) and use the depth rod to find and mark the desired length on a straight section of ⅜″ copper tube (P1). (See Figure 12-12, next page.)

Mount the blade of your tubing cutter on the marked line and cut the tube. Ream out both ends so that there is no internal lip. Slide two alternately facing ⅜″ flare nuts (F6) onto the tube with the large openings pointing outward.

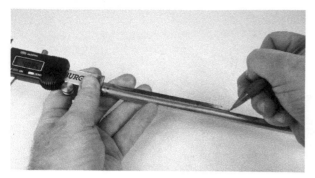

FIGURE 12-12: **Mark the tube.**

Position the end of the tube into the flaring anvil so that you will end up with a ⅛″ flare (see Chapter 9 for details of how to do this). Flare each end.

Complete the same set of measuring and flaring operations for the other tube.

Final Assembly

The reason that we've been measuring everything so carefully is that once we have the outside tubes complete, we have to mate the sections using only flare fittings. We've had to use swivel flare fittings because of the need to tighten sections without rotating them. The downside to this is that we have four fittings that all need to be within hundreds of an inch of fitting onto the receiving manifold. Even one of those fittings that is too long or too short will cause the others not to fit.

Flared fittings cannot tighten if the flare isn't flush to the male fitting without bending the flared rim of the copper tube. Bending like this weakens the flare and provides additional opportunities for leaks. Since flare fittings are not designed to be taped, if the flare doesn't serve to stop the escaping vapor, the threads will not do so. (See Figure 12-13.)

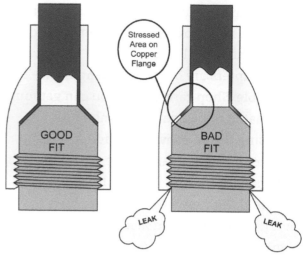

FIGURE 12-13: **Overstressed flare joint**

Disconnect the two central assemblies from the receive manifold. Attach the two newly fabricated tubes to their respective paths on the send manifold. Hold the path's flared ends up against the male flare fittings on the receive manifold. The female flared copper ends should fit flush against the male flare ends.

It's possible at this point that you will find that all of the paths from the send manifold are not parallel. You should be able to bend a path a few degrees to bring it to true. I've found that many ¼″ solenoids do not have their input and output paths in line with one another. The copper tubes you fabricated, or the swivel fittings of the central assemblies, will give you a couple degrees of movement to work with. But if you have to move them very far, you risk either abusing the flange or having to remeasure the outside paths and make new tubes.

It's better to make new tubes if you need them rather than overstress the ones you have. You'll find a use for them in the future as swivel adapters, so if you have to make up additional ones, take the long view and go ahead and do it.

With the flares mated from send to receive, begin tightening the flare nuts. Tighten each one a half to quarter turn in sequence so that they evenly pull the two manifold assemblies together without stressing any individual flare union. Tighten the flare nuts, being careful to hold wrenches on parts you don't want to rotate. (See Figure 12-14.)

FIGURE 12-14: Completed DSV

It is highly likely that you will have to do additional tightening while leak-testing (which is absolutely essential for this device). See the "Testing" section below for more detail.

Controller (Optional)

There are a variety of options for control of the DSV. Essentially, the goal is to open one of the solenoid valves. It may seem like you could get more resolution (pressure levels) by opening more than one valve at a time, but the pressure levels aren't additive; whichever of the open valves has the highest pressure will determine the output pressure. Unless you engage in advanced control approaches like pulse-width modulation (the details of which are beyond this text), opening one valve at a time is the goal.

The simplest approach is to have three switches, one switching 12V for each valve. Simply press the switch for the desired pressure level. Or you could have three outputs of a microcontroller serve as switches using MOSFETs. However, there are occasions, especially when using limited output pins on microcontrollers, where you would want to control the three valves with as few switches as possible, which in this case is two.

If we treat two switches as binary digits we can count to three; 00, 01, 10, 11. Since zero is included, this provides four states, which match the four lines of the DSV nicely.

SWITCH ONE	SWITCH TWO	SOLENOID SELECTED
Off	Off	No solenoid engaged, pilot pressure only
On	Off	First solenoid engaged, level one output
Off	On	Second solenoid engaged, level two output
On	On	Third solenoid engaged, full pressure

I wanted the DSV to be independent of the rest of my control systems, so I constructed a controller that takes two control lines and 12V DC as input and has control lines for the three solenoids as outputs. This worked nicely with four-pin trailer connectors. I really like trailer connectors for their availability, proven durability, and weather resistance.

Perhaps out of nostalgia, or perhaps because they also have proven durability, I chose to use the venerable 74xx TTL logic family to turn my two inputs into three outputs. A small, single-chip microcontroller is a viable option, and if you work with them, it's easy to duplicate the functionality of this circuit in software.

Converting two inputs into four states is known as decoding or demultiplexing. There are many chips that do this. The simplest of these is the 74x139 dual 2–4 decoder. The 74-series chips come in a dizzying array of variants: 74LS139, 74HC139, 74HCT139, 74AC139, 74ACT139, and so on. The distinctions between these aren't important to the discussion at hand; any version you can commonly acquire should work. I'm using a 74LS139 dual inline pin (DIP) package that fits in standard breadboards, protoboards, and sockets. (See Figure 12-15.)

The 74x139 chip has two 2–4 decoders in it. The DSV controller only uses one of them. I could use just part of a 3–8 decoder (74x138) if that had been the only chip available.

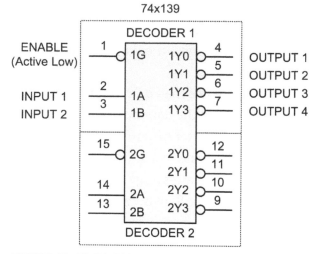

FIGURE 12-15: 74×139 diagram

Almost any decoder will work if you just use the number of inputs and outputs you want. For the 74×139, once the enable pin is active (which is done by connecting it to ground), the state of the input pins changes the state of the output pins as described in the following truth table.

You may have noticed that the output pins are normally high (1). This is because they are active low outputs. To change these outputs into the active high signals for use with MOSFETs we need to invert the signal. (See Chapter 11, "Arduino Control," for details on MOSFETs.)

INPUTS			OUTPUTS			
G (Enable Low)	A (Input 1)	B (Input 2)	Y0 (Output 1)	Y1 (Output 2)	Y2 (Output 3)	Y3 (Output 4)
1	N/A	N/A	1	1	1	1
0	0	0	0	1	1	1
0	0	1	1	0	1	1
0	1	0	1	1	0	1
0	1	1	1	1	1	0

To perform the inversion, we'll rely on another 74-series chip, the 74×04 hex inverter. This chip provides six inverters that automatically flip the incoming signal from high to low or low to high. By routing each of the outputs from the 74×139 through an inverter on the 74×04, we'll have the signals needed to drive MOSFETs.

Our circuit includes the 74×139, the 74×04, three MOSFETs, and flyback diodes. It also includes an On/Off switch and LED indicators to inform us if the circuit is powered and the state of the MOSFETs. Once again, we're operating 12V solenoids from a 12V source and need to power 5V components, so we'll include a 7805 voltage regulator to regulate the 12V down to 5V. (See Figure 12-16.)

You can build a circuit like this on any protoboard; I used a very basic board for this design. Screw terminals provide a means to attach the outboard components. I prefer to solder sockets rather than solder the chips directly to the board, but that's a matter of personal preference. Figure 12-17 shows the top-side layout.

FIGURE 12-17: **Component layout**

I used *point-to-point soldering* to connect the components. This method solders wires directly from component to component. In some cases, the components were placed so that I could use a solder bridge to join them to nearby pads. (See Figure 12-18, next page.)

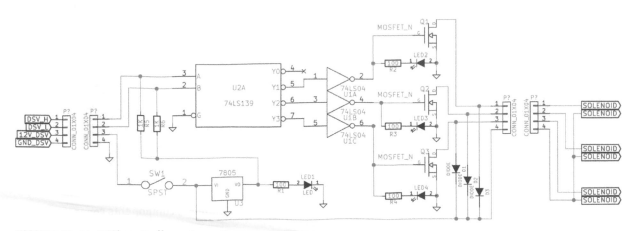

FIGURE 12-16: **DSV controller schematic**

FIGURE 12-18: **Soldered connections**

I wanted to build this controller to have durability and some degree of weather resistance, so I used a metal case. Grommeted holes like the ones we used in the boosh controller provide access for the four-pin trailer connectors (use a different gender for input and output) and regular holes allow mounting of the switch and LEDs. (See Figure 12-19.)

As I've described previously, it is a strong design aesthetic for me to always try to place all components on either the lid or the body of a case, but not on both. This allows for much easier troubleshooting. I typically squeeze silicone or hot glue (thermoplastic) into and around the holes to seal the wires, LEDs, or bolts. (See Figure 12-20.)

FIGURE 12-20: **The case lid**

Once the protoboard is mounted on standoffs (unconnected to electrical ground) the outboard components can be attached to the screw terminals. (See Figure 12-21.)

The solenoids are all wired to a four-pin trailer connecter with a shared positive input and connected to the controller. The way you

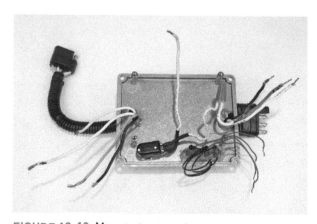

FIGURE 12-19: **Mounted external components**

FIGURE 12-21: **Outboard components connected**

wire the trailer connectors is arbitrary; Figure 12-22 shows how I did it.

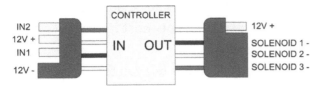

FIGURE 12-22: **Four-pin trailer connectors**

I mounted the controller case to one of the manifolds using 1″ split-ring hangers. (See Figure 12-23.)

FIGURE 12-23: **Mounts attached to the case**

Once attached to the manifold and connected to the trailer connector coming from the solenoids, the DSV forms a single unit with the controller. (See Figure 12-24.)

You still have to decide how you want to trigger the controller. Two switches or two outputs from a microcontroller will do the job; your choice will depend on how you want to use the device.

Burners

We've already referred to a number of options for burners. The flambeau, a simple straight burner; curved and fabricated approaches such as in the portable fire pit; or the pilot ring used in the boosh would all work with the DSV.

The only recommendations I can offer are regarding the size and length of the burner and piping. The DSV switches pressures quickly, but the pressure-driven vapor still has to propagate the length of the piping and the interior of the burner. If vapor from the high-pressure setting takes time to exit through the jets, this will slow the ability to switch to a lower pressure. So there are trade-offs. Generally speaking, short, narrow piping and large jets work the best with the DSV. Center or multipoint feeds also help evenly distribute the pressure wave to the jets.

FIGURE 12-24: **DSV and controller**

Testing

The full test plan for the DSV has to be constructed with the entire project taken into consideration. The most important test of the valve itself is leak-testing.

It would be less than honest if I didn't tell you that the DSV's major downside is leak-testing. I've constructed three of these, and every time, despite the lessons learned on the previous efforts, I've spent more time chasing down leaks than I did constructing the valve.

You can simplify leak-testing by testing the valve all by itself. The best way to do this is to place a ⅜" FFL cap on the output and connect the pressure source directly to the input. Your first attempts will likely detect leaks, so if you're in a location without excellent ventilation or that isn't free of ignition sources, you really should fabricate a pressurized air connection to use instead of propane.

Pressurized air can come from a tire, a compressor, a storage cylinder, or any number of other places. What's important for our use is how to get it connected to a ⅜" MFL connector. If you're using a compressor and have a compressor hose with a ¼" MIP end (which is common), you can simply add a ¼" FIP × ⅜" MFL adapter and a ⅜" FFL swivel adapter.

We did something similar with the water-tuning mechanism from the Venturi project, so if you built that device, you could even drain all the water out of the sprayer and hose, pump it up, and use it as a pressure source. If you're not sure how you'd get a pressurized source to hook up and test with, consider building that device.

We'll use the same leak-test kit described in all the plumbing chapters: a spray bottle filled with a small amount of dish soap. Make sure the DSV solenoids aren't hooked up to a power source, and spray all the DSV parts liberally. Don't worry if it gets sticky or slimy, because you can wash the soap off later.

Leaks at Taped Fittings

Threaded fittings leaks are both good news and bad news. The good news is that you can solve it by retaping and tightening the joint. The bad news is that you may have to completely tear down the DSV to do so.

Except for the fittings in the ends of the manifolds, all of the taped fittings in the four paths of the DSV are going to cause the solenoid and/or the needle valve to turn, as well. You might be able to get away with a needle valve turning (though it will likely throw it out of alignment with the other needle valves), but the solenoids will only turn until they contact the next path over.

My recommendation, upon seeing leaks in the taped fittings, is to smile, breathe deeply, and immediately tear the valve down as far as it takes to retape and successfully tighten the fitting.

Leaks at Flare Fittings

If you find leaks in the flare fittings, tighten them with one wrench on the offending fitting and one on the fitting it is attached to. Flare fittings aren't like taper thread fittings with tape. The flare fitting attempts to mate the male and female surfaces of the conical section around the orifice. This means that you can't just go on tightening forever.

If you're having to tighten the flare fittings on the swivel adapter or the sections of copper tube with flares we made up, know that overtightening will only result in stretching out (and ruining) the copper flare inside the flare nut. If a couple of quarter turns with the wrench don't do the job, you'll have to consider making up new, slightly longer, sections that will work properly.

Testing the Controller

Given the various options for switching the DSV, I can only offer generic tips, most of which are very similar to the troubleshooting in Chapter 11. Basically, test all your solder and wiring paths with a multimeter prior to powering the system. Once the system is powered, test the ability to activate the solenoids without having introduced propane yet.

If you encounter problems, check for bad solder joints and shorts. Test the output to the solenoids for 12V when activated, and work backward from there. If you're using salvaged trailer connectors or wiring, test them for continuity.

Operation

The DSV's operation is also dependent on your implementation. Ultimately, it should allow very fast switching of pressure levels. This is less impressive when you have long runs from the step valve to the burner. In that case, the line acts as an accumulator and slows down the change to lower pressures. Placing the DSV as close to the burner as possible helps this.

Once you have the DSV plumbed into your system and wired, close all the needle valves before providing pressure to it. Once these are closed, you can follow the steps in Chapter 4, "The Flambeau," to ignite the burner using only the pilot line needle valve. Start off using no more than 5–10 psi from the main regulator. Adjust the needle valve so that the burner flames are as low as possible without going out.

Using whatever switching mechanism you have derived, open the solenoid on the max pressure line (the one without a needle valve). Adjust the main regulator's output pressure so that the burner is at the level you want for the highest output. If you are using the DSV in a system that requires higher pressures from the main regulator for other effects, you will need to plumb a needle valve or additional regulator into the main input path.

Close the max pressure solenoid and engage the solenoid closest to the pilot path. Adjust its needle valve so that the burner is at your desired first-step level. Close that solenoid and do the same for the other central solenoid. This will provide four levels of output.

Ideas

This project was initially created as part of a multieffect mobile festival installation powered by a Raspberry Pi. The Raspberry Pi ran SuperCollider, a programming language designed for audio work. A shotgun microphone was attached to a USB audio input and routed to SuperCollider. As SuperCollider detected different levels of intensity in the music picked up from EDM dance camps, it would drive the DSV. The DSV had a large tube shaped as a square as a burner.

The project, called DeBoosh, had independent ¼" valve micro-booshes at each corner and a ¾" boosh in the center of the square. The heat from the main boosh pilot was used, with a radiator and water pump to warm a cylinder water bath. SuperCollider pulsed the flame curtain powered by the DSV in time with the beat of the music. When it detected interesting musical artifacts, like a bass drop, it fired off a pattern of the micro or main boosh. (See Figure 12-25.)

The DSV provided an interesting flame effect at a much lower pressure than the booshes. This allowed the effect to maintain visual activity without constantly freezing the cylinder supplying the booshes.

I hope that this project spurs your thinking. Whether you build the DSV, or use it as a jumping-off point for your own ideas, I want to encourage you to think beyond the basics of booshing. Enjoy!

FIGURE 12-25: DeBoosh project flame effects

Beyond Vapor

EVERYTHING WE'VE DISCUSSED IN this book has been about working with propane vapor. Indeed, since propane liquid isn't directly flammable, it turns to vapor in all but the coldest temperatures and must mix with oxygen; the only way to burn propane is as vapor. However, we have relied on the propane storage cylinder to convert the stored liquid into vapor for us. This is extremely convenient and allows us to use relatively low-cost components rated for vapor downstream from that cylinder.

Nevertheless, we've discussed the challenges in drawing significant amounts of vapor from a cylinder. The resulting temperature and pressure drop compromise the ability to continue to draw the volumes of vapor we might desire. This is an unacceptable design constraint in many propane systems. Forklifts, automobiles, and hot air balloons are all examples of propane-fueled systems that can't work for a while, and then sit around waiting for the cylinder to thaw out.

So how do these systems work? And, for the purposes of entertainment, are they viable for flame effects? This chapter will take a look at an individual maker who upgraded to a liquid propane effect and a world-class art collaborative that uses liquid in many large-scale projects.

Liquid Systems

Liquid propane flame effects tend to work in one of two manners: the first is by using a vaporizer to convert the liquid to vapor as close to the burn point as possible, and the second is to eject and ignite the liquid directly. Moving to a vaporizer system is the easiest first step from vapor-based effects. Direct liquid ejection can provide dramatic flame jets climbing 100′ (30 m) in the air, but are significantly more dangerous to build and operate.

Direct liquid ejection and ignition requires a very specialized skill set that I cannot begin to describe in this book. These effects are by far the most dramatic fire effects that most people will ever see. However, learning to work with liquid propane ejection is something that you need to learn from someone or some group with experience, like the Flaming Lotus Girls who we'll discuss later. For now, I want to concentrate on a viable next step for makers from vapor-based systems, like the ones described in this book—this is the use of vaporizers. The specific details of working with vaporizers are also beyond a text like this one, but I hope to provide some ideas and directions for folks interested in moving to this next step.

Vaporization

Vapor-based systems like the ones described in this book rely on the cylinder itself to act as a vaporizer. In Chapter 1, "Understanding Propane," we looked at how drawing vapor from the cylinder causes the liquid inside to boil and create more vapor. This is an easy model to work with, but has some frustrating drawbacks. The biggest one is that the more vapor you vent, the colder the cylinder gets. Staring at a cylinder with 2″ of ice around it that's putting out 15 psi is a lot less fun than making big fireballs.

Many fire artists will pursue some means of warming the cylinder to keep it from freezing. The options for doing this range from the safe to the insane. On the safe end are water baths and commercial tank heaters. On the insane end of the spectrum is any solution that uses flame or unregulated heat applied to a propane cylinder. Any solution that risks ignition or has the potential to inadvertently heat the cylinder above safe levels of pressure is unreasonably dangerous. Water baths can slow down the cooling by increasing thermal transfer, but unless the water is heated, you can freeze your cylinder in a tub of ice if you're not careful. Solutions that require electricity, such as commercial propane heater blankets, are a challenge to use out in fields and remote locations. *Never* use a heater blanket that isn't built specifically for propane cylinder use.

So, assuming the complexity of successful cylinder temperature regulation doesn't sound appealing, how do you get all the vapor you want out of cylinders without having them freeze over? This is a problem faced by a variety of uses from forklifts to industrial uses to hot air balloons. The answer is to move the primary vaporization out of the cylinder and into a specially built device, known as a vaporizer.

Since vaporizer-based systems move liquid from the cylinder, instead of vapor, different requirements for downstream components emerge. Drawing liquid from a cylinder and piping it through a system requires valves, hoses, and fittings that are rated for liquid use. Propane's corrosive properties are most potent when it's in its liquid state, so seals and gaskets that come in contact with liquid propane need additional ratings beyond vapor parts.

Liquid propane systems also have to take leak detection to an entirely new level. If a vapor system leaks, the primary risk is that the escaped vapor will ignite. Alternately, in an enclosed space, it could suffocate someone. If liquid leaks, we add the danger of frostbite, since it rapidly boils and chills far below the freezing point of water. High-pressure jets of freezing propane are a huge menace, and the resulting vapor cloud, generally far larger than all but the biggest vapor system leaks, is an extreme hazard for both fire and suffocation.

Nevertheless, the benefits a liquid propane vaporizer system offer for consistent delivery of propane still outweigh the risks, as long as proper equipment and processes are used. As of late 2014, there were over 143,000 propane-fueled vehicles (almost all use a liquid-based system) on the road in the United States, and this doesn't count the hundreds of thousands of propane lawn mowers and forklifts in use. Outside the United States, propane, sold as *autogas*, is the third most popular auto fuel in the world. The use of, and availability of, suitable parts for liquid propane is growing to the point that new and interesting opportunities for DIY liquid flame effects are increasing.

In a vaporizer system, the liquid is drawn from the cylinder either by a dip tube or by a valve below the level of the liquid. This is why forklift cylinders are on their side—so that the withdrawal valve can be below the level of the liquid. (See Figure 13-1.)

For vaporization to occur, heat is required. This is acquired either directly from the propane being drawn, which is burned to provide heat, or indirectly from steam, hot water, electricity, or another source of burning propane. The heat source boils the liquid into vapor in the vaporizer (sometimes called a converter).

Perhaps one of the most appealing sources of equipment for a liquid flame vaporizer effect is the world of hot air ballooning. Since 1960 when Paul (Ed) Yost

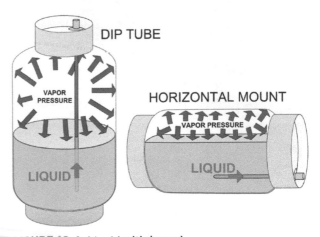

FIGURE 13-1: Liquid withdrawal

patented a successful propane burner that allowed balloons to carry their own fuel, hot air balloon enthusiasts have continually refined lightweight burners and equipment designed to make large open flames.

Hot air balloon burners are glorious external combustion devices. Beautiful pieces of engineering, they are actually fairly simple in concept. They use the heat of the output flame to warm and vaporize the incoming liquid propane. The liquid travels through tubes, which are wrapped in a spiral and situated in the path of the fire. (See Figure 13-2.)

When firing, these burners can be seen lighting up the night sky for miles. (See Figure 13-3.) Most balloons use a pair of burners, but four or more is not uncommon. A single burner can generally produce around 18 million BTUs (British thermal units) per hour. If run continuously, this would burn almost 200 gallons an hour, or just over 3 gallons a minute.

FIGURE 13-3: An ignited burner

Michael Boyd's Mega-Scorch

Balloon burners have been used by many makers as an upgrade to their flame effects. One interesting example is the Mega-Scorch dragon boat project designed by Michael Boyd. Mega-Scorch is a next-generation liquid propane boosh succeeding his original Scorch vapor project. Roaring fire from the dragon-head sculpture on the bow of a dragon boat is a great idea and Michael is a talented maker. His resulting build is a great project both on and off the water.

After having originally built a vapor-based, fire-breathing dragon as an entry in a

FIGURE 13-2: Hot air balloon burners

kinetic sculpture contest, Michael found that his ability to boosh was limited by tank chilling. Frustrated at the limitations of a vapor system, Michael acquired a second-hand T3-017 hot air balloon burner from eBay. It needed some minor repairs and would have cost a large amount of money to become certifiable as flight-worthy under Federal Aviation Administration standards. But Michael is a clever maker and wasn't looking to fly a balloon with the burner, so he was able to buy the burner cheap and get it in working order with the help of a local hot air balloon dealer. (See Figure 13-4.)

The goal wasn't just to create flames; Michael wanted to create a fire-breathing sculpture. A beautiful pair of plasma-cut, aluminum, dragon-head cutouts bracket the horizontally mounted burner, resulting in a truly awesome fire-breathing sculpture. Each side of the dragon head consists of two

cutouts to provide depth to the sculpture. Michael has constructed different mountings that allow him to use the effect on the boat as well as a stand-alone. (See Figure 13-5.)

FIGURE 13-5: Mega-Scorch burner in action
J.J. MAESTAS

Mega-Scorch is a tremendous crowd pleaser, lighting up at holidays and festivals. The system is easy to use, especially after Michael upgraded the stock manual valve used in balloons to a 12V electric solenoid valve. Having patiently waited, watching eBay and Craigslist for burner deals, the overall cost of the system was reasonable compared to trying to buy the equipment new or in flight-ready status. Michael is an intrepid maker and learned as he went with the project. The only big surprise for him was the tremendous amount of heat the burner put out, even using only the secondary burner (which puts out a "dirtier," more visible flame).

Stepping up from balloon burners rapidly brings us into the world of industrial class vaporizers. Effects that require continuously large quantities of propane vapor need cabinet-sized mechanisms that produce the vapor

FIGURE 13-4: Mega-Scorch burner test setup
DEBORAH J. BOYD

equivalent of hundreds of gallons of propane an hour. They are similar in output to a hot air balloon burner, but instead of flame as the output, you get a supply of vapor you can distribute. Flame effects that use devices like this can fire many dozens of booshes for hours at a stretch. But this industrial level of power is matched by industrial levels of cost. How can makers and enthusiasts take advantage of equipment like this by themselves?

The answer is to collaborate. The potential of pooled resources and experience is greater than the sum of the individuals. Much like upgrading from vapor to liquid, scaling up from individual maker efforts to group maker efforts can make truly dramatic things happen.

The Flaming Lotus Girls

One of the most exciting examples of the power of collaboration is the collective known as the Flaming Lotus Girls. A volunteer-based group founded in 2000 with six members that has grown to over 100 today, the Flaming Lotus Girls (or FLG) build large-scale "kinetic fire art." The girls (and all members, male or female, are referred to as girls) work in what's described as a "do-ocracy."

Do-ocracies are a fascinating organizational model. Essentially, they rely on people choosing to do things rather than being told what to do. This model may appear to have a whiff of anarchy about it, but it's surprising how effective it can be. FLG projects rely on someone in the group deciding that they will go do something. The group as a whole has no standing management or hierarchy. If a girl has an interesting idea, others may decide that they would like to do something to support it based on confidence, enthusiasm, or creativity.

People are welcome at any level of experience. Though FLG is oriented toward making more than teaching, they often hold classes on the skills necessary to get started in fire art. Nevertheless, new members must supply the most important ingredients in a do-ocracy—enthusiasm and motivation. It also doesn't hurt to show up with basic shop safety gear like work and nitrile gloves, safety glasses, hearing protection, and solid work boots. A basic tool bag with wrenches, ratchets, utility knife, duct tape, and the kinds of tools described in previous chapters demonstrates a readiness to contribute, also. (See Figure 13-6.)

The types of things going on in the FLG warehouse in San Francisco vary considerably depending on what's being built, so the opportunities to learn do as well. Big projects usually have multiple teams: electronics, plumbing, structure, and so on, and new members generally gravitate toward their area

FIGURE 13-6: FLG tools Caroline Miller (Mills)

of interest. Planning, organization, licensing, and other more informational tasks are also areas that new folks can get involved with.

Since many fire art activities have significant safety requirements, trust relationships are critical in an organization like this. Hanging around enough to listen, learn, ask questions, and generally be seen as committed is a key way to establish relationships with more experienced members. These relationships often resemble apprenticeships, in that time spent contributing to repetitive low-level work like grinding, cleaning, or labeling provides access to observe and learn more advanced skills. (See Figure 13-7.)

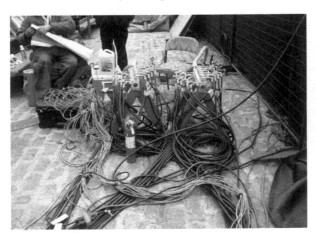

FIGURE 13-7: **Organizing supply lines**
CAROLINE MILLER (MILLS)

The skill set that exists among the FLG is impressive. Advanced welding and pipe fitting, fluid dynamics modelling, microcontroller circuit design, CNC, metal casting, and professional engineering experience are all examples of skills that the FLG can bring to bear on their large projects. Many of the skills required to make art at this scale are not necessarily intuitive, such as the need to be able

to design custom racks and assemblies to break down their monumental art so that it is shippable in one or more 40' shipping containers. When a piece like the Serpent Mother includes 91 individual vertebrae, each with an independent flame effect, a lot of work goes into the design and fabrication. (See Figure 13-8.)

The FLG fund projects through grants, fund-raisers, and donations. The girls are all volunteers, so the money goes to maintaining

FIGURE 13-8: **Serpent Mother vertebrae**
CAROLINE MILLER (MILLS)

FIGURE 13-9: **Serpent Mother spine and egg**
CAROLINE MILLER (MILLS)

their warehouse space in San Francisco, buying tools and parts, and storing and transporting projects. FLG work is often commissioned by festivals, cities, or art funding groups. Many of their projects cost tens to hundreds of thousands of dollars to create and display. (See Figure 13-9.)

The Flaming Lotus Girls build fire projects using propane vapor, propane liquid, kerosene, methanol, and various chemicals to color the flames. Many of their sculptures and installations combine a variety of flame effect types. A large FLG installation requires advanced logistical skills just to maintain a safe fuel depot for an event, let alone getting hundreds of parts, dozens of people, and all the supplies to an installation. (See Figure 13-10.)

FIGURE 13-10: **Flaming Lotus Girls fuel depot** CAROLINE MILLER (MILLS)

Industrial vaporizers are used in many FLG projects as a means to convert enough liquid propane to vapor for the dozens of poofers (I may call it a *boosh*, but the FLG calls them *poofers*) in a sculpture. Using industrial vaporizers liberates FLG from the problems of frozen cylinders and disappointing pressure drops. This allows consistency and reliability

at a level that direct vapor projects have a difficult time achieving. (See Figure 13-11.)

FIGURE 13-11: **FLG vaporizer** CAROLINE MILLER (MILLS)

Next Steps

Most makers who begin working with propane systems find that their needs never exceed vapor-based systems. It's been a recurrent theme in this book that I encourage flame artists to think outside the "bigger is better" arms race that often occurs among boosh enthusiasts. Liquid propane systems offer a means to increase the volume of vapor that is available, but that doesn't necessarily increase the creativity of the projects that use it.

While I wanted to give some idea of ways that liquid propane can be used, the next steps that I really want to encourage have to do with a different dimension of what Michael Boyd and the Flaming Lotus Girls have done. Michael Boyd added a dramatic new dimension to his dragon boat. Michael isn't just the coach of the Phantom Dragons dragon boat racing team; he races on Team USA in the Dragon Boat World Championships. He did it because he wanted to bring even more excitement to something he loved (and he's a fantastic maker!). The Flaming Lotus Girls create art that isn't just bigger, it's an expression of a group of people working toward a common vision. They use synergy to make art and the result, for the project and the people, is far greater than the sum of its parts.

If you've built a boosh and you're wondering what to do next, think about ways to take the excitement and beauty of fire, and safely place it in some new context. Combine it with something else you love to create something no one else has seen, or even thought of. Create patterns, make it sing, or make it move. Add sculpture to your structure, add air to sculpt the flames. Light a hundred small dancing fires instead of one great big one. Fire is a medium as old as humankind and we haven't even scraped the surface of what it can do artistically.

If you've built fire art and you're wondering how to do something more, look around, join a hacker space or a local burner community, and find other people who can merge their skills and energy with yours. I love my builds, but I love the people I've met and built with even more. There are few moments that are richer than sitting late into the night with a new friend, crafting art from pipes and valves, talking about everything and nothing as you learn to trust each other's skills. There are few more rewarding feelings than standing with your crew, exhausted from a huge build, seeing the piece you dreamed and worked for light up and stun an audience.

I hope this book provides you with a way to find your start in the world of fire art. There are many people who know much more about many aspects of this world than I do, and I hope that someday soon you'll become one of them. If you choose to express yourself with flame, I hope that you'll do so with your and other, safety foremost in your mind. I hope you'll share your work and art with the world, and I cannot tell you how much I look forward to seeing it.

Afterword:
Flame Effects and Burning Man

THE WORLD'S PREMIER FLAME effect showcase has got to be
Burning Man, the event attracting tens of thousands of people to
Nevada's Black Rock Desert every year. Dave X has been the fire
safety team manager for Burning Man for over 15 years. He is a
licensed pyrotechnic operator working in Northern California,
and along with his team, serves as the authority having jurisdiction (AHJ) for Burning Man. As the AHJ, the fire safety team
has reviewed and inspected thousands of effects over the years.
Dave X and his partner Eric Smith have taught close to a thou-
sand students in their flame effect classes in the United States
and around the world.

Nevertheless, Dave X's primary concern is the safety of the
audience and operators. I reached out to Dave X and asked him
to provide his perspective on fire art safety. He was incredibly
helpful and provided the following guidance. To help understand
the standards of bringing a flame effect to Burning Man, the event
has graciously allowed us to reprint the Burning Man Flame Effect
Guidelines, which can be found at http://burningman.org/event/
art-performance/fire-art-guidelines/flame-effects.

The Responsibilities of Those Who Work with Fire
and the Consequences of Their Negligence

by Dave X

In choosing to work with fire, we have chosen to work with a medium that implies a greater need for responsibility and safety than working with almost any other creative medium. If you spill paint, you clean it up with a rag; it will not burn the building to the ground. If you drop your camera during a photo shoot and crack a lens, no one is killed or maimed.

In my years working with fire and pyrotechnics, I have found there to be four areas of responsibility that must be addressed by those choosing to work in this field. If a creator fails their responsibilities in any of these areas, there are possible extreme consequences.

The Four Responsibilities of Those Who Work with Fire

1. Responsibility to Yourself

Fire burns.

Burns hurt!

Burns scar!

Most everyone has gotten a burn at some time in their life and can remember how much it hurt. Burns are one of the most painful injuries one can suffer. The pain of a burn lasts well beyond the time of injury and can last as long as a lifetime in the case of severe burns. Burn scars can be disfiguring and last forever. We have all seen severely burned folks and no one would choose to go through that for a lifetime just for a moment of fun.

Do Not Burn Yourself!

2. Responsibility to Others and Property

As much as burns can hurt you, they can hurt others…

Imagine the horror and guilt of burning or killing another person.

Small fires can build quickly to larger fires, burning rooms, buildings, blocks, and acres in very short order. We, as fire artists and performers, have chosen to work with this elemental danger and we accept the risks involved. While you may know the details of your piece and the warning signs of a possible accident, others do not. If property or person is burned by your work, even due to the negligence of others, you are responsible, because you brought it and represented it as safe.

Do Not Burn Others!

3. Responsibility to the Event or Venue Where You Display Your Work

The use of fire as an element in art, performance, and all types of projects has grown by leaps, and can now be seen worldwide. Those who open their venues for our use and those come to view your works are under the assumption you know what you are doing and their property and person are safe and secure. A single significant accident could cause ruin and suffering to those involved, due to lawsuits, liability, and the pain to the local community that an accident can cause.

Do Not Ruin It for Those Who Invite Us In.

4. Responsibility to the Fire Arts

As this new art grows, fire artists are now finding their work being sought out and included in larger festivals, as well as private and public art settings worldwide. Each of these types of settings has unique permitting and insurance requirements, and artists must now work closely with local fire authorities both before and during the exhibition of their work. The record of the safe use of flame effects is so far a good one, and this has gone a long way with those who review and permit this type of thing. If word spreads of injury, property damage, or death due to the use of fire in these type of projects, it might destroy years of work spent nurturing and legitimizing this new art form.

Do Not Ruin It for All Fire Artists!

—Dave X, 2016

Burning Man Flame Effects Guidelines

(reprinted with permission)

The primal simplicity of an open fire is great and all, but newfangled technology enables all sorts of spinning, swirling, squealing, pink-and-green fire magic, and Burning Man artists make full use of it. The thing is, flame effects involve lots of moving parts and high-pressure flammable fuels, so follow these guidelines to make sure you're doing it right.

Flame Effect Definition

Flame effect is defined as: "The combustion of solids, liquids, or gases to produce thermal, physical, visual, or audible phenomena before an audience." This includes all flames that are automated, switched, pressurized, or have any other action than simply being lit, as well as projects using propane or other liquid or gaseous fuels.

Safety Guidelines for Flame Effects

The majority of flame effects at Burning Man are liquefied petroleum gas (LP-Gas) effects; LP-Gas is often commonly referred to as propane. Most of the guidelines below deal with LP-Gas as a fuel. Regardless of fuel type or technological basis, all flame effects must be constructed in such a way as to meet or exceed applicable laws, codes, and industry standards.

The National Fire Prevention Association (NFPA) publishes numerous codes and standards for the construction and use of LP-Gas systems, including:

- NFPA 54—*National Fuel Gas Code*
- NFPA 58—*Liquefied Petroleum Gas Code*

- NFPA 160—*Standard for the Use of Flame Effects Before an Audience*

NFPA documents are available for viewing and purchase on the NFPA website and should be reviewed by all flame effects artists.

Construction of Flame Effects

- All LP-Gas cylinders shall be designed, fabricated, tested, and marked in accordance with the regulations of the US Department of Transportation (DOT) or the ASME *Boiler and Pressure Vessel Code*.

- All LP-Gas cylinders must have an unexpired certification date stamp and be in good working order. Tanks in poor condition or out-of-date are a danger to fill and may cause injury to the fuel team, the artists, and/or participants.

- Each LP-Gas flame effect must have a single quarter turn shut-off valve as the primary emergency fuel shut-off. When closed, this valve must inhibit all fuel flow to the flame effect, regardless of how many LP-Gas cylinders are connected to the flame effect. This valve must be exposed and visible at all times, and must be clearly marked as the emergency fuel shut-off.

- All components of the fuel system (fittings, piping, valves, connectors, etc.) must be designed and rated for both the type and pressure of fuel being used. The use of improper fittings can lead to leaks and failures in the fuel system, resulting in fires and/or injury.

- All LP Gas metallic piping and fittings that will operate at a pressure greater than 125 psi shall be schedule 80 or heavier.

- All LP-Gas hoses that will be operated in excess of 5 psi shall be designed for a working pressure of at least 350 psi and shall be continuously marked by the manufacturer to indicate its maximum operating pressure and compatibility with LP-Gas.

- Air or pneumatic line is not acceptable as fuel hose. LP-Gas degrades rubber hose not specifically designed for use with that fuel. This results in the hose cracking from the inside out, potentially leading to a catastrophic failure.

- Hose clamps are prohibited on LP-Gas hose at any pressure. All fuel hose connections shall be factory made, or constructed with a crimped fitting specifically designed for that purpose. Hose clamps are well known for cutting and chafing fuel lines or coming loose, possibly leading to catastrophic failure.

- All metallic tubing joints shall use flare fittings. The use of compression fittings or lead-soldered fittings is prohibited.

- Accumulators, surge tanks, and other pressure vessels in the system shall be designed, manufactured, and tested in accordance with the ASME *Boiler and Pressure Vessel Code* or the Department of Transportation (DOT) for the pressure of the gas in use.

- Any welding alteration of pressure vessels, or alteration or fabrication of other system components that hold pressure, must be performed by an American Society of Mechanical Engineers (ASME) certified welder, and must be stamped and certified as such.

- If the fuel supply pressure exceeds the maximum allowable operating pressure (MAOP) of an accumulator or other pressure vessel, a regulator shall be installed between the fuel supply and the pressure vessel to reduce the pressure below the pressure vessel's MAOP. A pressure relief valve shall also be installed in the pressure vessel, with a start-to-leak setting at or below the MAOP and a rate of discharge that exceeds the maximum flow rate of the supply container.

- Fuel tanks for stationary flame effects must be protected from vehicle traffic and be well-illuminated at night.

- Flame effects should be constructed and sited in such a way that the flame head and/or hot components are at least 6″ from the playa surface, to prevent baking or scarring of the playa.

- Any artwork, towers, or other structures that incorporate flame effects should be secured from the wind and encircled with an appropriate safety perimeter to prevent injury to participants.

Operation of Flame Effects

FLAME EFFECT OPERATORS Flame effects operators and assistants must be 21 years of age or older and be trained in the use of fire extinguishers.

Operators and assistants must wear fire-resistant clothing while operating flame effects.

PERSONAL RESPONSIBILITY No carelessness, negligence, or unsafe conditions with flame effects shall be tolerated. Do not drink, take drugs, or smoke when working with flame effects.

SAFETY PERIMETER An appropriate audience safety perimeter (and performer's safety zone, if applicable) shall be established well in advance of flame effects operation, and must be approved by FAST. Because of the variety of artwork that incorporates flame effects, a member of FAST will help you determine the correct perimeter distance.

In any case, a 20′ zone around the flame effects must be kept free of all combustible or flammable materials, and nothing should overhang this zone.

FUELING Only people familiar with the safety considerations and hazards involved are permitted to connect/disconnect LP-Gas tanks, or to do liquid fuel filling. Wearing personal safety gear (glasses, gloves, etc.) during liquid fuel filling is required.

DAILY SAFETY CHECK A daily safety check of all flame effect components and connections is mandatory before operation begins. Never start operation of a flame effect until the daily safety check is completed. If a safety hazard is identified either during the safety check or during operation, the fire safety liaison must delay or halt operation until the hazard is corrected.

OPERATING GUIDELINES Never light a flame effect until all performers, safety monitors, and participants are in place and ready.

Never operate a flame effect in such a way that it poses a danger to people or property.

ATTENDING TO FLAME EFFECTS Flame effects must never be left unattended. The winds in the desert are highly variable, and may create havoc in a poorly monitored installation. Any flame effect found running unattended will be shut down. Egregious and/or repeat offenses will result in the confiscation and/or disabling of the effect.

NO SMOKING OR OPEN FLAME ABSOLUTELY no smoking or open flame within 10' of any storage area where flammable liquids or fuel gases are stored. All fuel and flammables must be stored in approved containers, which must remain closed except when filling or dispensing, or when connected to a system for use.

MATERIAL SAFETY DATA SHEETS MSDSs for any hazardous chemicals used in the construction or operation of the flame effect must be kept at the installation, so they are available to guide cleanup activities in case of a material spill, and to provide to emergency medical personnel in case of accidental exposure.

The Gas Laws

BEGINNING IN THE SEVENTEENTH century, experimenters and investigators documented the physical properties of gases. These laws describe "ideal gases" rather than real gases, but real gases perform so closely to ideal gases that the equations are valid for all but the most extreme cases.

The three most important gas laws (at least for the purposes of this book) describe the relationship between the temperature, pressure, and volume of a gas. There are additional gas laws, but these three will serve to provide a foundation for understanding the physical properties of propane described in this book. Additionally, there is a fourth combined gas law that that merges the three primary laws into a one statement.

Boyle's Law

FIGURE A-1: **Robert Boyle**

Robert Boyle published the first of these laws in 1662. Boyle's law describes how the pressure of a gas decreases as the volume increases (and vice versa). This formula assumes constant temperature for the gas.

$$PV = k_1$$

or

$$P_1V_1 = P_2V_2$$

where

P = pressure of the gas

V = volume of the gas

k_1 = the constant described by the relationship between pressure and volume

Charles's Law

FIGURE A-2: **Jacques Charles**

Joseph Louis Gay-Lussac published the second law in 1787 but credited Jacques Charles (for whom it is named). Charles's law states that for gas at a constant pressure, the volume is directly proportional to the temperature (in a closed system). If the volume increases, so does the temperature (and vice versa).

$$V/T = k_2$$

or

$$V_1 / T_1 = V_2 / T_2$$

where

V = volume of the gas

T = temperature of the gas

k_2 = the constant described by the relationship between volume and temperature (different than the constant derived in Boyle's law)

Gay-Lussac's Law

In 1809, Joseph Louis Gay-Lussac published the third of the major gas laws. Gay-Lussac's law describes that, for a given mass of gas at a constant volume, the pressure in the container is proportional to its temperature. If the pressure increases, so does the temperature (and vice versa).

$$P / T = k_3$$

or

$$P_1 / T_1 = P_2 / T_2$$

where

P = pressure of the gas

T = temperature of the gas

k_3 = the constant described by the relationship between pressure and temperature (different than the constant derived in Boyle's or Charles's laws).

FIGURE A-3: Joseph Louis Gay-Lussac

Combined Gas Law

This law is the result of combining the previous three laws into a single statement. It states the ratio between temperature and the product of pressure and volume is a constant.

$$PV/T = k_4$$

or

$$P_1V_1 / T_1 = P_2V_2 / T_2$$

where

P = pressure of the gas

V = volume of the gas

T = temperature of the gas

k_4 = the constant described by the ratio between temperature and the pressure-volume product

Resources

THE FOLLOWING RESOURCES ARE by no means the only sources of reference or vendor information. The books are some of my favorites to get started with. Unfortunately, the URLs suffer from the vagaries of the Internet and may change or disappear over time.

Books

Banzi, Massimo and Michael Shiloh. *Getting Started with Arduino: The Open Source Electronics Prototyping Platform (Make), 3rd Edition.* San Francisco, CA: Maker Media Inc., 2014.

Gurstelle, William. *The Practical Pyromaniac.* Chicago, IL: Chicago Review Press, 2011.

Scherz, Paul and Simon Monk. *Practical Electronics for Inventors, 3rd Edition.* New York, NY: McGraw-Hill Education TAB, 2013.

Propane Art

"Dance Dance Immolation." *Interpretive Arson.* http://www.interpretivearson.com/projects/ddi/

Department of Spontaneous Combustion. http://www.spontaneousfire.com

El Pulpo Mecanico. http://www.elpulpomecanico.com

Flaming Lotus Girls. http://flaminglotus.com

"Pyrotechnics." *Chris Marion.* http://www.chrismarion.net/index.php?option=com_content&view=category&layout=blog&id=43&Itemid=226

Reference Information

"An investigation of Rubens flame tube resonances." *Brigham Young University.* December 2008. https://www.physics.byu.edu/download/publication/645

"Arduino – Home." *Arduino.* https://www.arduino.cc

"Electronics Primer: How to Solder Electronic Components." *Science Buddies.* http://www.sciencebuddies.org/science-fair-projects/project_ideas/Elec_primer-solder.shtml#materials

"Flame Effects Guidelines." *Burning Man Project.* http://burning-man.org/event/art-performance/fire-art-guidelines/flame-effects/

"Forge and Burner Design Page #1." *Golden Age Forge.* https://www.abana.org/ronreil/design1.shtml

National Fire Protection Association. http://www.nfpa.org

National Propane Gas Association. https://www.npga.org

"Procedures for Brazing Pipe and Tubing." *The Harris Products Group.* http://www.harrisproductsgroup.com/en/Expert-Advice/tech-tips/procedures-for-brazing-pipe-and-tubing.aspx

Propane 101. http://www.propane101.com

Propane Education and Research Council. http://www.propane.com

"Proper Brazing Procedure." *Lucas-Milhaupt, Inc.* http://www.lucasmilhaupt.com/en-US/brazingfundamentals/properbrazingprocedure/

"Rubens' Tube - Waves of Fire." *The Naked Scientists, University of Cambridge.* http://www.thenakedscientists.com/HTML/experiments/exp/rubens-tube/

Vendors

AdaFruit. https://www.adafruit.com

Adventures in Homebrewing. http://www.homebrewing.org

Amazon.com. http://www.amazon.com

Best Materials http://www.bestmaterials.com

Crystal-Technica. http://crystaltechnica.com

Harbor Freight Tools. http://www.harborfreight.com

The Home Depot. http://www.homedepot.com

Lowe's. http://www.lowes.com

Maker Shed. http://www.makershed.com

Northern Tool + Equipment. http://www.northerntool.com

PlumbersStock. https://www.plumbersstock.com

PlumbingSupply.com. http://www.plumbingsupply.com

SparkFun. https://www.sparkfun.com

Tejas Smokers. http://www.tejassmokers.com

Zoro. http://www.zoro.com

Index

Tweco tip, using with Venturi burner, 173
type M tubing, failed bends in, 101

U

UEL (upper explosive limit), 15

V

valve protection, 22
valves
    ball type, 38–39
    bibb type, 40
    buying, 36
    for cylinders, 36–37
    gate type, 40
    manual, 35
    needle type, 39–40
    plastic versus brass, 49
    quick connect, 38
    solenoid type, 40–42
    water type, 40
vapor pressure of propane, 9–10
vaporization, 308–310
vaporizer for FLG (Flaming Lotus Girls), 314
Venturi burner
    assembling, 168–169
    assembling jet, 167–168
    attaching choke, 169–171
    block diagram, 161
    chokes, 186
    cutting threads, 162–163
    dowel, fitting, and pipe, 180
    drilling holes, 165–167
    enhancements, 177–187
    flame characteristics, 175–176
    flares, 177–185
    flaring frame, 181
    flow, 186–187
    igniting, 175
    operation, 177
    parts, 160–162
    project overview, 158–159
    schematic, 160
    shims under tube, 174
    tapping cap, 164–165
    testing, 174–176
    tools, 162
    troubleshooting, 173–174
    tuning and testing, 171–176
    tuning jet, 172–173
    in tuyere, 176
    Tweco tip, 173
    water tuning harness, 171
    water tuning harness parts, 160
    zones of action, 187
vise, 54
voltage divider, 261

W

water tuning harness, constructing, 171–172
water valves, 40
waves in Rubens tube, 124–126
wire cutters, using for portable fire pit, 98
wiring test for Arduino control, 276–277
WOG valve rating, 35–36
working with fire, responsibilities
    related to, 318–320. *See also* fire
    extinguishers; propane
wrenches. *See* crescent wrenches; pipe
    wrenches

9 781680 450873